ANSWERS TO QUESTIONS PEOPLE ARE ACTUALLY ASKING

BOOK 2

ANDY FLETCHER

© **2021 Andy Fletcher**

Andy Fletcher
Answers to Questions People Are Actually Asking, Book 2
All rights reserved. No part of this book may be reproduced in any form or by any electronic or mechanical means, including information storage and retrieval systems, without permission in writing from the publisher, except by a reviewer, who may quote brief passages in a review.
Published by fletchpub at Lulu
Cover Design by: Lulu and Andy Fletcher
A CIP record for this book is available from the Library of Congress Cataloging-in-Publication Data
ISBN-13: 978-1-365-04893-7
Distributed by: Lulu

TABLEOFCONTENTS

5..Grüezi
7...........What some of the big words mean
9......................................Science Parts
117..................................God Parts
157..................................End Parts
159..........................Glossary of Terms

ANSWERS TO QUESTIONS PEOPLE ARE ACTUALLY ASKING

GRÜEZI MITENAND

That means "Howdy Y'all" in Swissgerman. I'm trying to get it to go viral. It's one of the world's great greetings.

Anyway.

This is the second book in a series of, well, two at this point, called Answers to Questions that People Are Actually Asking. You may have noticed it on the cover and the title page.

You don't have to have had read the first one to read this one. It's not a novel. It's a bunch of questions and answers.

In the last intro in the first book, I described how I spend a lot of time answering questions on a website called Quora, and these are some of those questions and answers. I've become a most-read writer in a dozen or so topics, mostly science topics, but also, dare I say it, some on God, faith, and some random bits of other things that just intrigued me.

I was pretty well convinced that only nerds and geeks go to Quora. Those are my people. My tribe. My homeys.

OK, seriously, old white guys should never say "homeys". Not even sure I spelled it right.

Anyway. I was lecturing on some of this stuff at a school in California, where, by definition, even the nerds are cooler than cool Midwestern guys, and when I casually told them I was a most-read writer on Quora, they, like, gave me an ovation, like it really meant something.

Who knew?

So here you go, more answers to more questions that people asked on Quora that I dared to answer.

Are these the right answers to those questions?

Don't be ridiculous. These are my answers. Some are right, some are sort of right, all of them are kinda right, none are completely wrong, and they are all meant to get a conversation going.

Read a little. Talk a little. Amongst yourselves, I mean. Don't bother me. I've got things to do.

Oh. Some of the questions are not written in the finest of English grammar. That's because some of the questioners are not native speakers. So cut them some slack. I could have fixed it, but I thought it was kinda charming. So I didn't.

ANSWERS TO QUESTIONS PEOPLE ARE ACTUALLY ASKING

WHAT SOME OF THE BIG WORDS MEAN

Actually, it's more like the anagrams:

MRW—Most Read Writer
NSFW—Not Suitable For Work
QM—Quantum Mechanics
GTR—General Theory of Relativity
STR—Special Theory of Relativity
BB—Big Bang
K—Kelvin
CMB—Cosmic Microwave Background Radiation
COBE, WMAP—Two of three experiments to measure the CMB
BH—Black Hole
EH—Event Horizon of a BH (the point of no return)
ST—String Theory
ST—Spacetime
IT—Inflationary Theory
MWI—the Many Worlds Hypothesis of QM
AP—Anthropic Principle
SAP—Strong Anthropic Principle
BTW—By the way
WAPO—Washington Post
QED—it has been demonstrated
GOD—God
LY—Light year
Byo—billion years old
Bya—billion years ago
DE—Dark Energy
DM—Dark Matter
SF, WF, EM—Strong Force, Weak Force, Electromagnetic Force
YEC—Young Earth Creationists
SOL—Speed of light
ETH—Einstein's university in Zurich
MOND—Modified Newtonian Dynamics (tweaking the theory of gravity)
M-theory—a version of String Theory
CCC—an hypothesis of Big Bangs repeating over and over (Roger Penrose)
NIST—National Institute of Standards and Technology (Boulder)
TV—television (JK)
JK—Just Kidding

ANSWERS TO QUESTIONS PEOPLE ARE ACTUALLY ASKING

SCIENCE PARTS

CAN REALITY CHANGE?

Sure, it changes all the time. The question is, how would you know? If you're talking about your own personal reality, then change is what it does, constantly and unremittingly.

If you're talking about reality at large, that is, the reality of the universe at the level of the laws and constants of nature, then those constants don't change. If any of them did, then order, structure and life itself go away, so we don't want that to happen, do we?

Does your reality changing alter the nature of reality at large? That's a Chaos Theory question, but where that holds a chance of being true on the planet, off the planet, not so much, at least not yet. The Butterfly Effect tells us that small changes can sometimes have huge impacts, but those impacts are still restricted to the planet.

Climate change is a good example—a 1% of 1% change in the composition of the atmosphere has led to a 1% (and growing) change in the temperature of the Earth's oceans, but it's having a large impact on the entire Earth, in a Chaotic, unpredictable sort of way, an impact that will get larger and larger.

But that's on the planet. Off planet, no change.

IS THE FACT THAT THERE ARE EXPERIMENTS SHOWING THAT PARTICLES CAN ACTUALLY BE ENTANGLED THROUGH TIME? LIKE IN THE 'DELAYED CHOICE EXPERIMENT' AS AN EXAMPLE?

Here you go, from Aeon.com:

Up to today, most experiments have tested entanglement … across space. But what if entanglement also occurs across time?

Just when you thought quantum mechanics couldn't get any weirder, a team of physicists at the Hebrew University of Jerusalem reported in 2013 that they had successfully entangled photons that never coexisted.

Previous experiments had already showed quantum correlations across time, by delaying the measurement of one of the coexisting entangled particles; but Eli Megidish and his collaborators were the first to show entanglement between photons whose lifespans did not overlap at all.

IF THE EVENT HORIZON OF BLACK HOLE IS ALWAYS IN THE FUTURE OF ANYTHING FALLING TOWARDS IT, HOW CAN IT BE THAT 2 BLACK HOLES ULTIMATELY CAN MERGE?

Though there is a lot of disagreement on BHs, time and space, it seems that from our perspective, the merger would never be complete, only almost complete forever, or at least until the new BH evaporates via Hawking radiation. It takes an infinite amount of time for a BH to become a BH, so there are no completed BHs, just a lot of almost BHs, and they'll keep getting closer and closer in a Zeno's paradox sort of way until they evaporate. Evidence?

"By merging two seemingly conflicting theories, Laura Mersini-Houghton, a physics professor at UNC-Chapel Hill, has proven mathematically that black holes can never come into being in the first place. The work not only forces scientists to reimagine the fabric of spacetime, but also rethink the origins of the universe." (phys.org - Sept 2014)

"...all the information that fell down the hole was actually trapped on the black hole's two-dimensional event horizon, the surface that marks the point of no return. The horizon encoded everything inside, like a hologram. It was as if the bits needed to re-create your house and everything in it could fit on the walls." (quantamagazine.org)

"...Stephen Hawking's final paper has just been published, revisiting the question of whether information can be retrieved from a black hole or whether it is lost forever. The paper posits that information can be stored in a halo of 'soft hair' surrounding a black hole." (Science | Ars Technica)

WILL THE COSMIC FRONTIER BE OUR HOME SOME DAY OR WILL WE NEVER LEAVE THE SOLAR SYSTEM?

Hate to be a nay-sayer, but I'll say, nay. We're never going to leave. Humans, that is. As far as sending Voyager and Pioneer ships out of the Solar System, they're just in empty space and will travel through empty space forever, unless they get hit by something solid and speedy.

Sending a ship to the nearest star, which has planets, would require a minimum of 99 people to start with, and it would take 6300 years to get there. That's 300+ generations of people who will be born, live and die on the space ship having made no decision to be there and with no way to get off. By the time their descendants arrive, they will have developed new language, new culture, new religions, will have no memory or meaningful history of Earth or anything on it, and may not even know what a planet is. And then the planets they find there are unlikely to be habitable. Oops. Now what do we do?

We're never going to leave. Might as well take care of Earth a little better. It's all we are ever going to have.

WHAT WOULD BE STRANGER: IF OUR GALAXY WAS THE ONLY ONE, OR IF THERE ARE INFINITE GALAXIES?

If our galaxy were the only one we could see (to tweak your question into a more realistic one), then we'd have no access to evidence for Big Bang, so we would not know that the universe had an origin. We could conceivably look for and find the CMB, but we'd have no reason to do so. The GTR would have told us that the universe was expanding, but as it is currently only expanding between galactic clusters and not inside them, esp not inside ours, we would think that the GTR was wrong, and who knows what that would have done to our subsequent use of it? We might have rejected it as valid out of hand, and thus we might not have GPS locator systems, which have to rely on both Special and General Relativity to be accurate.

Since the universe is expanding more rapidly, at some point in the far distant future this will be the case— we will be unable to see any other galaxies, assuming we are still around, and thus we will no longer have evidence of the universe having had a starting point. We will be the only galaxy as far as we will ever know, and we will then be likely to assume that the universe is infinitely old and empty.

It's worth noting that everything we now understand about the universe has come from the startling discovery that it had a Big Bang at the beginning, that it is not infinite in time, and maybe not in space.

Since we do not know whether or not it is infinite in space, if it is, then there would be an infinite number of galaxies already. So that's not strange;

it's possible. Einstein didn't think it was infinite in size, so I'm gonna go with Albert on this one.

WHAT CAN ONE TELL ME ABOUT WHERE INFLATION CAME FROM AT THE FIRST MOMENT OF THE BIG BANG? WHAT IS IT MADE OF? WHERE HAS IT GONE? HOW DOES IT WORK. I.E., HOW IT INFLATES?

Though Alan Guth and Andrei Linde, two of the architects of Inflation, could no doubt have more to offer than I, as far as I know nobody knows where Inflation came from, where it has gone, or how or why it worked at the time that it did. What is it made of? A theoretical particle/field called the Inflaton, about which we know nothing.

The origin of Inflation was to find a way to explain how the universe was such a low entropy place back at the beginning, how the temperature and density varied by only one part in 100,000 even in places too far apart for that information to have been communicated even at the speed of light.

The only way for that to have happened was for the universe to have been very hot, dense, and with all of its parts extremely close together, and then for it to have expanded unbelievably rapidly for an unbelievably short amount of time. One figure that I have seen says that it expanded from the size of a proton to the size of a grapefruit; another from the same starting point, the size of a proton, to 250 million light year across; but in both cases, it happened in a trillionth of a trillionth of a trillionth of a second—10^{-35} of a second.

When that happened, the temperature and density were spread like paint over the entire 4D surface of the universe incredibly evenly, to that one part in 100,000, even though the parts of the universe were now many light years away from each other.

So something like Inflation had to have happened. Apart from that, and the predictions that Inflation makes that are true, we don't know very much more about it.

WHAT IS THE ESSENCE OF QUANTUM MECHANICS, IN AT MOST THREE SENTENCES?

Here's my shot at it. I'm going to use semi-colons. It's a cheat, but I can live with it.

Particles can be in 2, 3, many, or all places at the same time, or nowhere at all, or anywhere at all; particles communicate with each other faster than light and backwards in time; particles go from here to there without going anywhere in between; particles appear and disappear out of and into nothing without a cause.

But nothing ever happens without an observation or measurement; thus particles don't exist until they are observed; in other words, human subjectivity is drawing forth the world. Without us, there is no reality.

We have 100% control over reality existing, and 0% control over what that reality will be.

WHAT IS THE BIGGEST QUESTION OF OUR UNIVERSE?

You can ask the universe any question you like, but the universe is not an answering thing. You can investigate it using math and physics and find answers to questions that we seek—why is there something rather than nothing, why did the universe do anything at all after Big Bang, where did the laws of physics come from, why is there order and structure, is life mandated or accidental, are observers really a critical need for reality to exist—but the universe itself neither listens nor answers.

If you want to rephrase the question, assume (if only briefly, if that matters) that God exists, then things get more interesting: Are You there and do You care? That would be my top vote getter. My personal view is that both sides to that question have been asked and answered to the affirmative, which then leads to a host of fascinating questions about good and evil, suffering, the meaning of life, what the heck is religion and who decided that was a good idea, and tell me more about these Jesus guy. But that's just me.

WHAT ARE THE STRANGE QUANTUM FORCES THAT ENABLE A GECKO'S FEET TO STICK TO WALLS?

It's called the Van der Waals force. From several articles:

"Researchers pondered for years how the gecko, which is substantially heavier than a spider, was able to scamper across ceilings so easily. Biologist Robert J. Full of UC Berkeley showed that the clinging ability was produced by minute hairs on the animal's feet. Each gecko toe contains more than 100,000 tiny hairs, and each of those hairs is split at the ends into hundreds of even tinier tips.

"The individual hairs get so close to whatever surface the gecko is climbing on that they generate a Van der Waals' force, the same force that holds molecules together. The force generated by each tip is vanishingly small, but added up, it gives the gecko tremendous clinging ability... (source unknown)

"...research led by Kellar Autumn, one of the nation's leading experts on gecko biomechanics, revealed for the first time that a gecko keeps its feet sticky but clean by shedding dirt particles with every step.

'It goes completely against our everyday experience with sticky tapes, which are 'magnets' for dirt and can't be reused,' said Fearing..." (source unknown)

"MIT researchers have created a waterproof adhesive bandage inspired by gecko lizards for patching up surgical wounds or internal injuries.

"Drawing on some of the principles that make gecko paws unique, the surface of the bandage has the same kind of nanoscale hills and valleys that allow the lizards to cling to walls and ceilings..." (ScienceDaily Feb. 19, 2008)

"... Researchers have developed a directional adhesive, inspired by the gecko.

"The polymer fibers are 600 nanometers in diameter, just 1/100 the diameter of a human hair.

"Like the gecko, the synthetic microfiber array is not sticky except when fibers slide a small distance along a surface.

"What sets this new gecko-inspired adhesive apart from the others created thus far is that it is directional, only "sticking" when it slides along a smooth surface, not when it is pressed down.

"'A gecko running uphill may be attaching and detaching its feet 20 times a second, so it'd get very tired if it had to work hard to pull its feet off at every step.'" (Source unknown)

WHAT'S THE MOST INTERESTING EXPLANATION YOU CAN COME UP WITH FOR

THE EXISTENCE OF OUR UNIVERSE (NOT NECESSARILY A SCIENTIFICALLY ACCURATE ONE, OF COURSE)?

Frankly, Big Bang is by far and away the most interesting. The evidence we have collected over the past 9 or so decades (!) is that we have a singular Big Bang producing a unique (as in, there's only one) universe out of essentially nothing in a trillionth of a trillionth of a trillionth of a second, and that was only the Inflationary Epoch. In far less than one second, an entire cosmos popped into being, and not just a cosmos, but all of the laws of physics, space, time, energy, matter, and the potential for everything else just bounced into being for no apparent reason, since there was no science to cause it to happen yet. The one possible exception is quantum mechanics, but that doesn't help, because it's just an assumption that QM was operating and caused the universe, and it is becoming more and more clear that QM absolutely mandates an observer.

And then the universe arrived with almost zero entropy, that is, in the highest state of order (thermodynamic equilibrium) that it would ever be in, and then this order produced chaos, which then produced order, structure, and ultimately you and me.

AND the universe arrived with over 200 constants of nature that have to have almost the exact values they have in order for order to arrive—an unbelievably level of fine-tuning.

If you had tried to tell the world that this was true before we had discovered any of it, you would have been hooted from the room as a religious nut.

None of the many other alternatives that have been proposed comes close, neither in how interesting they are, or in having any evidence to support them. Big Bang is so interesting that science still struggles to accept it (because it sounds too much like religion) and many religious people struggle to accept it (because it sounds too much like science.)

ARE RELATIVITY AND QUANTUM MECHANICS SACRED WORDS? ARE YOU EMOTIONAL ABOUT THEM?

They are breathtaking.

Particles can be in 2, 3, many, or all places at the same time, or nowhere at all, or anywhere at all.

Particles communicate with each other faster than light and backwards in time.

Particles go from here to there without going anywhere in between.

Particles appear and disappear out of and into nothing without a cause.

But nothing ever happens without an observation or measurement

Thus particles don't exist until they are observed.

In other words, human subjectivity is drawing forth the world. Without us, there is no reality.

We have 100% control over reality existing, and 0% control over what that reality will be.

From Roger Penrose at Oxford:

"Quantum reality is strange in many ways. Individual quantum particles can, at one time, be in two different places—or three, or four, or spread out throughout some region, perhaps wiggling around like a wave. Indeed, the "reality" that quantum theory seems to be telling us to believe in is so far removed from what we are used to that many quantum theorists would tell us to abandon the very notion of reality when considering phenomena at the scale of particles, atoms or even molecules.

"This seems rather hard to take, especially when we are also told that quantum behaviour rules all phenomena, and that even large-scale objects, being built from quantum ingredients, are themselves subject to the same quantum rules. Where does quantum non-reality leave off and the physical reality that we actually seem to experience begin to take over? Present-day quantum theory has no satisfactory answer to this question. My own viewpoint concerning this—and there are many other viewpoints—is that present-day quantum theory is not quite right, and that as the objects under consideration get more massive then the principles of Einstein's general relativity begin to clash with those of quantum mechanics, and a notion of reality that is more in accordance with our experiences will begin to emerge. The reader should be warned, however: quantum mechanics as it stands has no accepted observational evidence against it, and all such modifications remain speculative. Moreover, even general relativity, involving as it does the idea of a curved spacetime, itself diverges from the notions of reality we are used to.

"Whether we look at the universe at the quantum scale or across the vast distances over which the effects of general relativity become clear, then, the common-sense reality of chairs, tables and other material things would seem to dissolve away, to be replaced by a deeper reality inhabiting the world of mathematics."

They are only sacred if something sacred had the definitive role in creating them and organizing the universe around them. I would lean in that direction:

"A growing number of people think that what really matters are not things but the relations in which those things stand.

"...we may never know the real natures of things but only how they are related to one another.

"Now a question arises: What is the reason that we can know only the relations among things and not the things themselves?

"The straightforward answer is that relations are all there is." (Scientific American)

"On the other side are quantum physicists, marveling at the strange fact that quantum systems don't seem to be definite objects localized in space until we come along to observe them—whether we are conscious humans or inanimate measuring devices.

"Experiment after experiment has shown—defying common sense—that if we assume that the particles that make up ordinary objects have an objective, observer-independent existence, we get the wrong answers.

"The central lesson of quantum physics is clear:

"There are no public objects sitting out there in some preexisting space.

"As the physicist John Wheeler put it, 'Useful as it is under ordinary circumstances to say that the world exists 'out there' independent of us, that view can no longer be upheld.'" (Quanta Mag April 2016)

To sum up—the quantum universe in which we live is not real in the way we perceive it to be. First, it mandates observers, and we seem to be them. Second, that act of observation is just one more interaction in a cosmos defined by interactions, by relations between things that aren't really there except as by-products of interactions.

It is only logical to realize that not only is the universe trundling along via interactions, but it arrived via an interaction, that Big Bang is a product of an interaction, a relation, a relationship, an observation.

And that ... is sacred.

I READ THAT CARL SAGAN ONCE SAID THAT IF THERE WAS ONE LESS CELL IN THE

UNIVERSE, IT WOULD COLLAPSE. DO YOU BELIEVE THAT? WHY OR WHY NOT?

I recently read that if there were one fewer hydrogen atom per cubic meter in the universe (on average), or one more (on average), then there would be no life, order or structure possible. The number needed, I read, is 5 per cubic meter to ensure that things are as they are.

That's quite a bit different from what you said Sagan said, but it makes a lot more sense, since that adds or subtracts 20% from the total mass of the universe, and as small as it seems per cubic meter, it's a crapload.

DO SCIENTISTS ACCEPT THE IDEA OF OTHER UNIVERSES?

Many do, many don't. There's no evidence for even one other universe, and the many who don't say that there never will be any evidence. The many who do say there will be evidence. The Don'ts are winning so far.

The three theories (using the term very loosely) that suggest other universes are String Theory (for which there is no evidence), Inflation (which is not true yet, though it has made predictions) and the Many Worlds Hypothesis of QM, which is and forever will remain just an hypothesis. Recent work indicates that the many many other universes predicted by String Theory are probably not possible, and the same is true for Inflation.

So a much better question would be, why do any scientists at all accept the idea of other universes without any evidence or any hope for evidence?

It's really because of Fine Tuning. Our universe is incredibly finely tuned to produce and support intelligent life, so much so that even atheists like Stephen Weinberg and Richard Dawkins admit that without a multiverse, a benevolent designer (Weinberg's words to Dawkins) is the only other possible solution.

So many of the many who do, do because it's the only alternative to God, and God, to them, cannot be an answer.

For some of the many who don't, God has become an alternative.

WE MADE UP MATH BUT WE DIDN'T MAKE THE UNIVERSE, THEN WHY IS MATH = THE UNIVERSE?

Math is the language the universe speaks. It is written in math, understood only via math. We discover the language, are discovering the language, but we don't speak it perfectly, and in some cases we don't speak it at all yet. Sometimes we spoke it badly or got it completely wrong.

But we don't make up math. We make up the grammar, discarding it when it doesn't work, keeping it when it does. We make up the symbols, but they are only symbols, useful in translating the reality of the universe into a form we can understand.

But the language is there, waiting to be discovered, waiting for us to learn how to speak it, how to write it, how to put it together into coherent sentences.

We bumble along in the darkness, the universe babbles along in mystery, and we learn how to speak to each other.

WHY DOES THE UNIVERSE SEEM TO BE AS IT IS?

Your question is more interesting than you think, because the way the universe seems to be via our perceptions is nothing like the way that it really is.

Take Special Relativity. Time goes slower the faster you go, which is not perceptible to us because the change is so small at the speeds we have access to. Space also compresses as you go faster. At the speed of light, all of time and space are compressed into a single instant in two dimensions. Essentially, at the SOL, everything happens at once and in the same place. For a photon, the universe is a plane, even a point, where events are not separated by time.

Take General Relativity. Gravity is an interaction between matter and spacetime, matter telling spacetime how to curve, spacetime telling matter how to move (with thanks to John Wheeler for the phrase). Time passes more slowly closer to the center of gravity, so that time is passing more slowly at the bottom of your feet than the top of your head, again the difference being so small that you can't tell. "Down" is a time word much more than a space word; things that fall, do so by moving in the direction in which time is passing more slowly.

Take Quantum Mechanics. Particles can be in two, three, or all places at the same time. Particles really aren't anywhere, in fact, until they have been observed. Observation determines reality. Particles can go from here to there without traveling the space in between, and do so instantaneously. Entangled particles communicate when observed at faster than the speed of light.

Particles seem to know when they are being observed. Particles can be in multiple states at the same time, and only become real, one state remaining while all the others go away, when observed. Everything we call real is made of things that cannot be regarded as real (thanks to Niels Bohr for the phrase). In fact, everything real seems to be real only via interactions -there's nothing really in the universe but interactions. There are no objects in spacetime, just interactions. Everything seems to be here via interactions—space, time, gravity, the laws of physics, energy, matter, and maybe the universe itself. And interactions need observers, as observation seems to be the defining and ultimate interaction that creates reality.

Take Big Bang, out of General Relativity. Everything came from nothing in a tiny fraction of a second. Space, time, the laws of physics, energy, matter. Matter just pops into existence out of energy. Energy pops into existence out of nothing. The universe was not here, no space, no time, no laws of physics, and then it was. And nothing happens without an observation, without an observer. Not even the universe itself.

Take Fine Tuning. There are over 200 constants of nature (the gravitational constant, the values of the Strong, Weak, and Electromagnetic forces, the value of the Higgs Boson and the Cosmological Constant, to list the most important among many others), the values of which are so carefully dialed in that if you changed any one of them by a tiny amount, order, structure, life and even the universe itself go away.

Why? Why are all of these things the way they are?

God only knows.

HOW IS THE UNIVERSE MADE FROM THE BEGINNING?

Big Bang. Out of nothing, or nearly nothing, space, time, the laws of physics, energy and simple matter popped into being, not caused by any law of physics (since they didn't exist yet). The new universe had almost zero entropy (disorder), almost perfectly ordered with regard to temperature (it was nearly the same everywhere) and density (ditto), the same to 1 part in 100,000. Quarks, gluons, and electrons popped into existence out of the background energy, along with their anti-matter equivalents, and 1 out of every billion or 10 billion times, the particle would arrive without its anti-particle, and matter came into being because of that tiny asymmetry.

So two tiny asymmetries (temperature/density and particles/anti-particles), along with the laws of physics and the brand spanking new spacetime, made everything happen—simple atoms, gas clouds, stars, nucleosynthesis, elements, more stars, more elements, galaxies, planets, life, complex life, and finally Elvis. And all the rest of us.

So the laws of physics in low entropy space and time made everything happen because of two tiny asymmetries and, of course, fine-tuning, which means all the constants of nature were incredibly balanced to make it all happen. But they all came into existence after BB.

We don't know why. God, maybe. It's as good a guess as any, better than most.

WHY ARE THE HUMANS THE CENTER OF INTEREST IN THE UNIVERSE?

Because according to quantum theory (dating back to the very beginning in the 1920s and recently validated to an extremely high precision of accuracy) and fine-tuning (that is, the universe has precisely the values it needs to produce and sustain intelligent life), humans are necessary for the universe, indeed for reality itself to exist as intelligent, sentient, self-aware beings who observe reality into being. Your basic Copernicans and Newtonians don't like that at all, because it violates the assumption that humans don't matter at all, but New Scientist magazine tells us that there never has been any evidence that this extreme version of Copernicanism is true, and in fact the evidence shows dramatically that it is not. Plus the universe is not Newtonian—it is relativistic and quantum, not to mention Chaotic and Complex. Nothing happens without an observation. Nothing. Not even the universe. Not even reality. As astronomer Alan Dressler has said, "We've abandoned the old belief that humanity is at the physical center of the universe, but must come back to believing we are at the center of meaning." Evidence on request.

WHAT IS THE CHANCE THAT A UNIVERSE SO FINELY TUNED AS OURS COULD MAKE ITSELF IN ONLY 13.8 BILLION YEARS?

In order for anything to have happened in the universe at all (something rather than nothing), the universe had to arrive after Big Bang in the lowest state of entropy that it would ever be in, a state where variations in temperature and density varied only by 1 part in 100,000.

The Odds against a low entropic early universe happening by random chance...

...are 10 to the 10 to the 123rd to one, against. That's from Roger Penrose, Rouse Ball Professor of Mathematics at the University of Oxford.

It's a very very big number. How big?

Penrose: "Even if we were to write a `0' on each separate proton and on each separate neutron in the entire universe-and we could throw in all the other particles as well for good measure-we should fall far short of writing down the figure needed."

How far short?

If you wrote a zero on every particle in the observable universe, you would need to do the same inside this many universes:

Ten trillion trillion trillion trillion trillion trillion trillion trillion trillion trillion additional universes.

Just to write the number down.

If you tried to write it down, you'd have to program a computer to write 100 billion zeros a second and let it run for a hundred million trillion trillion trillion times the age of our universe.

The odds are non-existent that the universe is ordered by random chance.

The odds against life happening by chance are greater than that.

So life cannot have happened by chance. Anywhere in the universe. Not just here. Anywhere.

And that's just entropy.

The level of fine-tuning is so extraordinary that even atheists are impressed—Stephen Weinberg told Richard Dawkins that the only choices were a benevolent creator or a multiverse. He of course prefers the multiverse, but according to Lee Smolin, Roberto Unger, John Horgan, Margaret Wertheim, Peter Woit, and even Stephen Hawking and Thomas Hertog in Hawkings' last paper, the multiverse is not an option, is even pseudoscience, untestable, without any possibility of ever being supported by a scintilla of evidence.

So other opinions to the contrary, the chance of universe being finely-tuned is zero, effectively, and the only option is God or the Multiverse, the latter of which is far more a claim to unsubstantiated faith than the former, for which fine-tuning stands as pretty compelling evidence. Other opinions seem to be driven by a distaste of having to consider God as an option in a scientific context rather than by the science itself. The same was true, of course, for Big Bang, originally rejected because it sounded too much like religion, now kind of accepted as long as we don't talk about the G-word, and for which there are no other possible options that are supported by evidence.

IF ANOTHER VERSION OF YOU CAME AND VISITED YOU, AND TOLD YOU THAT THEY ARE YOU FROM A PARALLEL UNIVERSE, WHAT WOULD THEY HAVE TO HAVE TO SAY TO CONVINCE YOU THEY ARE TELLING THE TRUTH?

Not a dang thing. A DNA test showing identical genes would be interesting, but it would be more likely for someone to have cloned me than for other universes to exist. There are 3 hypotheses that seem to allow for multiple universes, but recent work on two of them (String Theory and Inflationary Theory) also seem to eliminate most if not all other universes. And there will never be any evidence that the Many Worlds Hypothesis of Quantum Theory is either true or false. Not to mention that there's never going to be any way even to detect other universe, much less to travel between them. So it's gonna be cloning, a dream, an hallucination, excellent make-up, a TV spoof (look around for the hidden cameras!), a Nigerian scam (don't give your clone your ATM card!), Halloween (pay no attention to the ax sticking out of other-you's head), Deja-You, or some other much more logical, reasonable explanation.

ARE THE MAJORITY OF THE "STARS" THAT WE SEE IN THE SKY ACTUALLY FULL GALAXIES, BECAUSE OF HOW INCONCEIVABLY VAST THE UNIVERSE IS?

I have read that on the darkest of nights including both hemispheres, we can see at most 9110 stars with the naked eye. And they're all stars.

And in other pictures of the observable universe, the pinpoints of light are not stars, not galaxies, not clusters of galaxies, but super-clusters of galaxies, that is, clusters of clusters of galaxies. It's really, really big.

And I saw estimates that the rest of the universe, the unobservable universe, is maybe $10\wedge10\wedge30$th times bigger than our part, and maybe $10\wedge10\wedge10\wedge122$nd times bigger. To put those numbers into some kind of perspective, it is estimated that there are 10^{30} microbial cells on Earth, which is far more than the estimated number of stars in the observable universe, and that there are 10^{120} elementary particles in that same universe.

So we see essentially nothing.

WHAT ARE THE SIMILARITIES OF THE BIG BANG THEORY AND STEADY STATE?

Big Bang is the actual history of the origin of the universe, all of the evidence points towards BB, and no evidence points to any other theory. Steady State was Fred Hoyle's alternative of choice, and there is no evidence in support of it; all the evidence points against it. Opinions against BB and/or for SS are just that—opinions not based in science or evidence. If you want evidence for BB, check out some of my answers on this site.

BTW, BB was rejected by many scientists, including Fred Hoyle, because it sounded just like the first chapter of Genesis. So if you are a person of faith, BB is lovely. If you are not, it is still lovely and entirely consistent with a God who made it happen.

WHAT ARE THE THREE STAGES THAT ESTABLISHED THE UNIVERSE?

1. Singularity
2. Big Bang and/or Inflation, not necessarily in that order and Inflation isn't science yet.
3. Everything else.

IF THE BIG BANG IS THE CREATOR OF TIME, THEN COULD THE OBSERVER EXIST WITHOUT IT? IF SO/NOT, THEN WHY?

Time and space did not exist, and then, with BB, they did. Observers the way that we understand them (and everything) exist by occupying points in space and time. So no observer as we would understand them could have existed at Singularity, before time began.

But quantum theory increasingly mandates an intelligent observer.

And everyone seems to agree that BB was a quantum fluctuation that needs some sort of mechanism to measure the Singularity and hence the universe into existence.

But before time and space began, there were no mechanisms. No observers as we might understand them.

John Wheeler suggests that human observers collapsed the wave function backwards in time to cause the universe to come into being. Which means that a universe that wasn't here yet somehow produced humans.

God seems like a more likely solution to the problem. Science says that God can never be an answer to a problem in the universe. The laws of physics will fill in all those gaps.

Certainly true. Inside the universe. Inside time and space.

But since the laws of physics arrived in the universe along with spacetime, then we have no laws of physics to have caused BB except for QM. Which mandates an observer.

So once again, God seems like the most likely solution to the problem. Whether or not one likes that is more a matter of personal preference rather than a measure of the evidence.

WHAT IS THE REAL-LIFE APPLICATION FOR THE BIG BANG THEORY?

The idea that the universe had a beginning (the BB) came from the math of the General Theory, which in turn started with Special Relativity. Both STR and GTR are used every time you crank up your GPS to find your way to anywhere you are trying to get to. That's an indirect application.

A direct app? When we stopped believing that the universe and the laws of physics were always here, infinitely old and large, we suddenly had to answer questions that didn't even exist—where did the universe come from? why is there something rather than nothing? why did the universe do anything? why is there order and structure? where did the laws of physics come from? And the answers to those questions have changed the way that we understand and approach everything in science.

It even has reintroduced the idea of God into the conversation. That's a real life application of note.

ARE THERE ANY ABSURD THEORIES ABOUT WELL KNOWN STORIES THAT ACTUALLY MAKE SENSE?

In terms of science, relativity (STR and GTR) and quantum theory are absolutely absurd, make no sense at all, and are perfectly true. That may not be exactly what you had in mind, and it's kind of the opposite of what you requested, but there you go.

Relativity (STR)—the faster you go, the slower time passes, but not for you, just relative to wherever you left, and the more space compresses towards two dimensions, but again, not for you but relative to wherever you left. Space and time are really spacetime, which is what the universe is made of. GTR—spacetime warps and bends under the presence of anything made of matter, i.e. has mass. Time passes more slowly closer to the center of gravity, so time passes more slowly at your feet than at your head. And it just gets weirder from there.

Quantum Theory—as weird as weird gets. Particles can be in 2 or 3 or many places at once, can go from here to there without traveling anywhere in between can communicate (!) instantaneously far faster than the speed of light (!), can skip regions of space entirely, can be teleported instantaneously, and aren't really particles—they're just "temporary ripples in an energy field." What they are depends upon how you are looking at them, and reality is thus determined by making a measurement or an observation, which means that observers are necessary for the universe itself to exist. And it just gets weirder from there.

ARE HUMAN BEINGS GOVERNED BY CLASSICAL MECHANICS OR BY QUANTUM MECHANICS?

From David Deutsch in Discover Mag:

"That quantum theory is outlandish, everyone agrees,' says Deutsch. It seems completely in conflict with the world of big physics according to Newton and Einstein.

"'To grapple with the contradictions, most physicists have chosen an easy way out: They restrict the validity of quantum theory to the subatomic world.

"But Deutsch argues that the theory's laws must hold at every level of reality.

"Because everything in the world, including ourselves, is made of these particles, and because quantum theory has proved infallible in every conceivable experiment, the same weird quantum rules must apply to us."

ARE THERE RULES FOR THE COLONIZATION OF THE UNIVERSE?

Let's hope to God there are. Otherwise, when they get here to colonize Earth, it's going to get ugly.

CAN THE UNIVERSE BE A SCIENTIFIC EXPERIMENT OF ADVANCE ALIEN CIVILIZATIONS AND WE THINK WHAT THEY MADE US TO THINK?

Why would they allow you to figure this out, then? Are they ready to call the experiment off or finished? It's quitting time in the alien lab? Time for a long alien weekend? Some counter-revolutionary alien has snuck into the lab and let you know about the experiment, and now the rest of them are in a blind panic ("They know! They know! OMalienG! We gotta fix this!) and hunting him down as he hides in the ceiling above the alien urinals? It's a glitch in the alien code? Some alien Russian or Ukrainian hackers have snuck a bug into the code, and now you know about the experiment? The mind reels. Or maybe we've reached the point where we have become so advanced that they are now ready to talk to us, though Trump and Brexit would seem to mandate against that.

But clearly the aliens are telling me what to think, so I wouldn't trust me at all. I wouldn't even trust me when I tell you not to trust me. And don't send me all your money via Paypal. Unless you really want to screw with the aliens. Then I'm happy to send you my Paypal thingy.

They wanted me to say that. Or not.

IS THIS UNIVERSE REALLY AS DULL AS IT SEEMS, LIFELESS AND DARK APART FROM THE STARS?

Only if you restrict yourself to looking at it at our speed. If you speed up the clock, then it's a dynamic and fascinating place, full of star birth and death, galaxies colliding, black holes devouring things, galaxies rushing apart, all kinds of cool stuff. It's just too slow for us to see it. That's not the universe's fault.

At the quantum level, everything is turbulent and wild. But it's too small for us to see.

Too big, too distant, too slow, too small. Not the universe's fault.

IF A PHOTON HAS ZERO MASS, HOW CAN IT BE TRAPPED IN THE GRAVITATIONAL FIELD OF A BLACK HOLE?

In order to move away from the black hole singularity, anything inside the event horizon has to travel back up a time slope, that is, backwards in time. The only way to go backwards in time (apart from that weird quantum thing) is to go faster than light, which light obviously can't do. The slope to the singularity is not a spatial slope, because the singularity no longer occupies a spatial location, just a temporal location. It doesn't exist in space anymore; just in time, and it's always in the future. So the only way out, is back up the time slope.

WHAT WOULD HAVE HAPPENED IF THE BIG BANG NEVER HAPPENED? WOULD WE EXIST?

BB caused the actual universe to come into existence. Without it, there's no universe, no space or time, no laws of physics, no energy or matter. There is nothing. Your question assumes that the universe was here and then BB produced... what? That's the wrong assumption. BB produced everything. So, no. No you or me, no empty space, no time. Nothing.

IS TIME A DIMENSION THAT CAN BE TRAVELED?

It depends upon what you mean by "travel." But as space and time are really spacetime, the "fabric of the cosmos" as Brian Greene put it, then we are always traveling through spacetime.

But even so, it seems that time is the more determinate direction. For example, "down" is defined as moving in the direction in which time passes more slowly, much more than moving down in space. "Down" is a time slope much more than a space slope. Time passes more slowly at the center of the Earth than on the surface (the center is 2.5 years younger than the surface, the same for the sun; 40,000 year difference between core and surface), and for you—your feet are slightly younger than your head.

When spacetime gets greatly exaggerated by gravity, as in or near a black hole, then the time slope tends to win. Outside the BH, if you are watching something trying to go to the event horizon, that object or person will disappear into the future much more than into space, heading down the time slope. If you are that person trying to hit the event horizon, if you go feet first, you will get spaghettified into the future, down the time slope, rather than in space (spaghettification seems to be more likely to happen at the event horizon for an object with length rather than inside the BH), and your feet will disappear into the future, along with, gradually, the rest of you, eventually shredded in time rather than space, your various bits distributed temporally rather than spatially.

Inside the BH, there is much more space by many orders of magnitude than you would expect. but you head down the time slope towards the singularity, which no longer has a spatial location, only a temporal one. The singularity is always in the future. The reason that nothing can escape a BH is that anything trying to do so has to go up the time slope, that is, backwards in time, and that's not possible for subliminal or even luminal objects.

However, quantum bits can be entangled in time, so it is possible for quantum particles to go backwards in time and might be possible thus to escape the BH. Maybe. That could be one way for BHs to evaporate, though I haven't seen anybody propose it. If it's brand new and if I thought of it, watch this space for the Nobel Prize. They can send the prize money in Bugattis.

WHAT IS THE BASIS OF SCIENTIFIC CRITICISM OF CREATIONISM?

There are a number of different bases for criticizing creationism from a science perspective. Useful, first, to differentiate between different types of creationism; young Earth (6000 years old), old Earth (it's, well, as old as it looks, something over 4 billion years), and intelligent design (ID) (which tends to be old Earth but tragically has been saddled with the word "design," which implies a blueprint of a universe that should be perfect (since God presumably is) but is radically not perfect (evil and suffering and all of that). Young Earthers are anti-evolution, anti-Big Bang. Old Earthers are pro-Big Bang but can be anti-evolution. ID-ers are pro-Big Bang and generally pro-evolution, but a designed, directed evolution rather than random and undirected, and also tend to insert a miraculous moment or two into the evolutionary process. These are all general tendencies and there are variants.

There are also a growing number of people of faith who accept evolution and Big Bang, but have a different view of how order and structure arrived in the universe apart from Intelligent Design or miraculous moments. More on that presently.

There have been significant moments in the co-history of science and faith where the faithful did not distinguish themselves. Geocentrism was one such moment; the Earth simply had to be the center of the universe. That's clearly not an issue any more for anybody of any faith, so that illustrates that faith can move on from those unfortunate passions based on faulty readings of scripture in a literal rather than in the poetic or figurative way in which it was originally written.

"God-of-the-gaps" was another; any time religious folk hit something for which they had no explanation and there was a gap between scientific periods of change, that gap would be filled with "God must have done it." Paul Davies' excellent book "God and the New Physics" deals with this well, though it is a couple of decades behind scientific progress at this point. The laws of physics ultimately cause everything to happen.

Evolution is a third, one that still divides many of the faithful from science. It's linked to a young Earth version of the universe; God miraculously and suddenly did everything—created the universe, Earth, life, man and woman, all in 7 literal days. The view of traditional evolutionary theory is that change happened gradually over long periods of time in a random, undirected way, driven by random genetic mutations that about 1% of the time turned out to match some environmental challenge and was reproducible, so that one individual with that mutation would cause it to spread and eventually take over, a new, mutated version dominating over the old.

And just to mention it, Flat Earthers are back, many of them people of faith, again using literal readings of what was meant to be poetic or figurative language to justify their belief that the Earth is really flat. Spoiler alert—it isn't.

The scientific critique of faith, however, is at its core based on two things; a Newtonian view of the universe and a Copernican view of the universe.

The Newtonian view is that the universe is infinite in time and space (which it is not, but that is implicitly assumed) and driven by the laws of physics which have also been here eternally (which they have not, but again is assumed.) That universe is deterministic, mechanistic, and reductionistic.

That is, by taking things apart to smaller and smaller parts (reductionism), one can discover which laws are making those parts interact in the way that they do and how they work together (mechanism), such that one can make predictions into the future and descriptions back into the past about why things have worked the way they have worked and how they'll continue to do so. So ultimately, everything is predetermined by the laws of physics and entirely predictable. There's no free will and no need for God.

This is mostly but not completely true. The universe is not purely Newtonian, but Einsteinian; relativistic and quantum. (It is also Chaotic and Complex, but that'll take more time and space than I want to take here.)

It is a universe that had a starting point of time and space, a point where the laws of physics came into being along with time and space, with energy and matter.

So the laws of physics have not always been here. Neither has the universe itself.

And that universe is quantum at its core; not predictable, but probabilistic, not deterministic, and not reductionistic, since at the smallest, quantum level, not only do we not understand things better, we don't understand things at all.

So free will is back, at least theoretically. And while the God of the Gaps argument holds true when you have the laws of physics to fill in the gaps, we have a time when the laws of physics didn't exist, and before that, nothing at all to have caused Big Bang to happen—no laws of physics apart from quantum theory, which requires an observer. So God might be back into the equation.

As far as the Copernican view goes, science made a similar, unjustifiable leap of illogic that religion did. Theologians had said that since the Earth and mankind are important to God, the Earth must be the spatial center of the universe. Wrong.

Scientists then said, since the Earth is not the spatial center of the universe, then it is not special, and neither is mankind. That is an assumption that is not based on facts in any way, and with the arrival of fine-tuning and the measurement problem of quantum theory, seems to be wrong entirely.

And evolutionary theorists, driven by the Copernican Principle, assumed randomness in evolution, and looked for randomness only. But with the arrival of epigenetics plus spontaneous emergence in Complexity Theory, and the return of Lamarckism, it seems more and more probable that evolutionary processes are driven by intent and specificity. Couple that with fine-tuning and the need for intelligent observers, and suddenly mankind matters and God has a role to play.

Finally, both science and faith are still bound into a Newtonian, Copernican view of the conflict between them, science lumping all religions into young Earth creationism, faith rejecting evolution and Big Bang. Many scientists, out of that same mold, refuse to consider that God is a reasonable answer to fine-tuning and the Observer problem, or even a reasonable answer to anything, still stuck on the God-of-the-Gaps objection without realizing that GOTGaps is dependent upon the laws of physics being eternal. And they are not. Many people of faith insist on literal interpretations of passages that are clearly full of a higher, deeper meaning.

And back to Flat Earthers, who insist that because we are so small and the Earth is so big that it looks flat, that it must be flat. People of faith read 7 days and insist on 7 24-hour days well before the sun and Earth had come into existence.

And scientists insist that because we are so small and the universe is so big, that mankind must be pointless.

The point? Gravity is the weakest of all the forces, so weak that we can't understand how. But it organizes the universe. Chaos Theory tells us that something as small a butterfly can flap its wings and wreak changes across the planet. So then, mankind, as the film Contact describes us, tiny and insignificant, but rare and precious. And God is the playwright, the architect, the choreographer, the poet, the painter, the sculptor.

IS REALITY REALLY REAL?

Great question. I mean, even just to look at it. And guaranteed to bring the nut jobs out of their caves. No offense to the nut jobs.

So I'll let the scientists themselves speak to the issue:

"Reality is but an illusion, albeit a persistent one." Einstein.

"At the most, it may be permissible to say that one can think of particles as more or less temporary entities within the wave field whose form and general behavior are nevertheless so clearly and sharply determined by the laws of waves that many processes take place as if these temporary entities were substantial permanent beings." Erwin Schroedinger—1952

"There is no such thing as reality, there is no such thing as free will, or things can travel faster than light." (Bell's Theorem—only the last one is true.)

"... experiments led by a group at the University of Vienna, Austria, provide the most compelling evidence yet that there is no objective reality beyond what we observe. Rather than passively observing it, we in fact create reality." (New Scientist mag, June 2007)

"The bizarre nature of reality as laid out by quantum theory has survived another test, with scientists performing a famous experiment and proving that reality does not exist until it is measured." (phys.org, May 2015)

"(In) April (2015), Nature Physics reported on a set of experiments showing a similar effect using helium atoms. Andrew Truscott, the Australian scientist who spearheaded the helium work, noted in Physics Today that '99.999 per cent of physicists would say that the measurement... brings the observable into reality.' In other words, human subjectivity is drawing forth the world." (Margaret Wertheim, aeon.com, Dec 2015)

"... a coherent description of reality, with all its quantum quirks, can arise from nothing more than random subjective experiences. It looks like the perspective of a madman,' because it compels us to abandon any notion of fundamental physical laws. 'It's deeply odd: you end up with a universe built directly from our experiences.'" (NewScientist Nov 2017)

"For almost a century, physicists have wondered whether the most counterintuitive predictions of quantum mechanics (QM) could actually be true.

"Only in recent years has the technology necessary for answering this question become accessible, enabling a string of experimental results—including startling ones reported in 2007 and 2010, and culminating now with a remarkable test reported in May—that show that key predictions of QM are indeed correct.

"Taken together, these experiments indicate that the everyday world we perceive does not exist until observed, which in turn suggests a primary role for mind in nature.

"Now that the most philosophically controversial predictions of QM have—finally—been experimentally confirmed without remaining loopholes, there are no excuses left for those who want to avoid confronting the implications of QM.

"Lest we continue to live according to a view of reality now known to be false, we must shift the cultural dialogue towards coming to grips with what nature is repeatedly telling us about herself." (Scientific American May 2018)

"A growing number of people think that what really matters are not things but the relations in which those things stand.

""…we may never know the real natures of things but only how they are related to one another.

Now a question arises: What is the reason that we can know only the relations among things and not the things themselves?

"The straightforward answer is that relations are all there is." (Scientific American, Dec 2015)

"As we go about our daily lives, we tend to assume that our perceptions—sights, sounds, textures, tastes—are an accurate portrayal of the real world.

"Sure, when we stop and think about it—or when we find ourselves fooled by a perceptual illusion—we realize with a jolt that what we perceive is never the world directly, but rather our brain's best guess at what that world is like, a kind of internal simulation of an external reality.

"Still, we bank on the fact that our simulation is a reasonably decent one.

"'Not so, says Donald D. Hoffman, a professor of cognitive science. Hoffman has spent the past three decades studying perception, artificial intelligence, evolutionary game theory and the brain, and his conclusion is a dramatic one:

"'The world presented to us by our perceptions is nothing like reality.

"'Getting at questions about the nature of reality, and disentangling the observer from the observed, is an endeavor that straddles the boundaries of neuroscience and fundamental physics.

"'(The) quantum physicists (marvel) at the strange fact that quantum systems don't seem to be definite objects localized in space until we come along to observe them.

"'Experiment after experiment has shown—defying common sense—that if we assume that the particles that make up ordinary objects have an objective, observer-independent existence, we get the wrong answers.

"'The central lesson of quantum physics is clear:

"'There are no public objects sitting out there in some preexisting space.

"'As the physicist John Wheeler put it, 'Useful as it is under ordinary circumstances to say that the world exists 'out there' independent of us, that view can no longer be upheld.'

"'… quantum physicists have to grapple with the mystery of how there can be anything but a first-person reality.

"'In short, all roads lead back to the observer.

"'Objective reality is just conscious agents, just points of view.

"'Interestingly, I can take two conscious agents and have them interact, and the mathematical structure of that interaction also satisfies the definition of a conscious agent.

"'... the idea that objectivity results from the fact that you and I can measure the same object in the exact same situation and get the same results —it's very clear from quantum mechanics that that idea has to go.

"'Physics tells us that there are no public physical objects.'"

"(AG) 'I suspect they're reacting to things like Roger Penrose and Stuart Hameroff's model, where you still have a physical brain, it's still sitting in space, but supposedly it's performing some quantum feat.

"'In contrast, you're saying, 'Look, quantum mechanics is telling us that we have to question the very notions of 'physical things' sitting in 'space.' '

"(DH) 'The neuroscientists are saying, 'We don't need to invoke those kind of quantum processes, we don't need quantum wave functions collapsing inside neurons, we can just use classical physics to describe processes in the brain.''

"'I'm emphasizing the larger lesson of quantum mechanics: Neurons, brains, space ... these are just symbols we use, they're not real.

"'It's not that there's a classical brain that does some quantum magic.

"'It's that there's no brain!

"'Quantum mechanics says that classical objects—including brains—don't exist.'" (Atlantic Monthly, April 2015)

"The third discovery ... is the most profound and difficult...

"The theory does not describe things as they 'are': it describes how things 'occur,' and how they 'interact with each other.' It doesn't describe where there is a particle, but how the particle shows itself to others. The world of existent things is reduced to a realm of possible interactions. Reality is reduced to interaction. Reality is reduced to relation.

"In a certain sense, this is just an extension of relativity...Quantum mechanics extends this relativity is a radical way: all variable aspects of an object exist only in relation to other objects. It is only in interactions that nature draws the world.

"... there is no reality except in the relations between physical systems. It isn't things that enter into relations, but rather relations that ground to the notion of thing.

"The world of QM is not a world of objects; it is a world of events. Things are built by the happening of elementary events.

"A stone is a vibration of quanta…just as a (ocean) wave maintains its identity for awhile before melting again into the sea.'" (Carlo Rovelli, *Reality is not what it seems*—2017)

So there you go. The reality that we experience is only one tiny slice of the reality that actually exists, so our perceptions lead us astray and allow us to assume that what we see is the way things actually are.

But the real reality, inaccessible to us perceptually, seems to be dependent upon our interactions with the universe in order for that reality, and indeed the tiny slice of reality that we occupy, to exist at all.

Ultimately, reality seems to all interactions. Reality is relational. The universe is a place defined fully by relationships, by interactions, and there is no other definition.

I'd be tempted to say that love is all you need. And that God is love. So maybe I will.

DOES CURRENT ASTROPHYSICAL RESEARCH CONCLUDE THAT THERE IS NO "EMPTY" SPACE IN THE UNIVERSE?

All empty space is filled with virtual energy and virtual particles doing this oscillation, this dance back and forth between energy and matter—$e=mc^2$, to be precise. It's an oscillation that goes from energy to particle/anti-particle and back to energy, each particle/anti-particle pair existing only for 10^{-21} of a second.

This is not astrophysical research, though. It's quantum mechanical. Space on average is empty at any one moment in time, but if you add up all the moments, it's full of quantum fluctuations, virtual particles and energy.

From Popular Mechanics Oct 2017: "…the uncertainty principle works for other quantities, too. The same principle applies to energy and time. The more you know about a particle's energy, the less you know about when it is, and vice versa. Here, something weird happens: If you know that there will never be a particle at a particular point, suddenly that point could have any amount of energy, sometimes enough to create a particle anyway.

"These particles are called 'virtual particles,' and they're basically quantum fluctuations. Once you make enough 'nothing,' the universe starts trying to find a way to fill it, even if that means creating particles out of thin air to do it."

This is also true inside of every proton and neutron, 99% of which are empty, only 1% of which is quarks. 99% of the mass of each is virtual particles, so that essentially, 99% of your mass is virtual. Add to that the fact that each atom has a million billion times more emptiness than somethingness, and none of us are really here, except for virtual particles, which also aren't really here, but in a persistent sort of way.

From Scientific American mag: "An extreme case of particles' being unpinpointable is the vacuum, which has paradoxical properties in quantum field theory. Look closely at any finite region of an overall vacuum—by definition, a zero-particle state—and you may observe something very different from a vacuum. In other words, your house can be totally empty even though you find particles all over the place...the theory predicts a particularly mind-boggling behavior of the vacuum: the average value of the number of particles is zero, yet the vacuum seethes with activity."

WHY DO PEOPLE ASK WHAT CAME BEFORE GOD, BUT THEY DON'T ASK WHAT CAME BEFORE THE BIG BANG?

It's because 1) people believe in BB and 2) don't believe in God and 3) don't understand BB and 4) therefore don't understand God, because 5) BB changes the whole conversation about God.

BB was at first rejected by science because it sounded too much like religion; everything coming from nothing for no reason in almost no time at all. They believed, at the time, that the universe was infinitely large and old and the laws of physics had always been there, and everything came from something, and that was because the laws of physics made everything happen.

So even now they make inane comments about God that reflect this. Like, who created God, or, what caused God? And so on.

But if they now accept that BB happened without a cause (as they tend to do), then clearly things can happen without a cause. And if they accept that BB produced everything from nothing, then clearly, something can come from nothing.

And if God was the architect of BB, then he doesn't need a cause or a source.

Determinism doesn't even happen in this universe, thanks to it being a quantum universe.

Being created is something that happens in this universe according to the laws of physics, which didn't exist prior to BB. God doesn't need to be created.

God, if he exists, does so outside of spacetime and the laws of physics. He doesn't exist like you and I do, occupying points in spacetime. He exists outside of space and time and the need to be created.

In fact, in the same way that you can't ask what came before BB (there was no "before" BB since time didn't exist yet), you can't ask what came before God. He's not a part of time or space.

Unless he chooses to be. So, as an aside, you can't find God. He's not there to be found. He has to find you, and that by entering the universe in a form you can recognize and, one would imagine, relate to. A 4D form, as it were. One example springs to mind.

WHAT ARE THE CORE REASONS FOR THE EXISTENCE OF BIOLOGICAL LIFE FORM IN THE UNIVERSE?

Science offers two answers to this question.

One, the Newtonian, Copernican answer is that there is no reason. Life is just an accidental by-product of random physical processes. There's no purpose for the universe nor any purpose for anything in the universe, including life, any form of life, and any particular person, like you or me. There's no reason for you to be here, no reason for me. Most science-oriented folks are going to go that way, even though they don't live like they believe it—most people need to believe that there's a purpose for them to be here in order to get out of bed in the morning. So they pretend, eventually forgetting that they are pretending, until they have to answer a question like this.

Two, the Relativistic, Quantum, fine-tuning answer is that humans are the observers who bring the universe into being simply by interacting with the universe. That's the Quantum bit. The fine-tuning bit is that the universe is astonishingly precisely calibrated to produce and support intelligent life, and we seem increasingly like the only intelligent life in the universe (that's still the fine-tuning bit). The Relativistic (Big Bang) bit is that the universe had a starting point, a place where none of those finely-tuned bits existed yet, likewise space and time.

The second answer is much better science than the first answer.

IF AN OBJECT TRAVELED FASTER THAN THE SPEED OF LIGHT, WOULD WE STILL SEE IT?

Technically, if something could go faster than light in a vacuum (which is not possible), it would be going backwards in time, so you'd better start now, because if it happens in the future, it should be passing by you in time any moment now. Oops. You missed it. Dang.

WHAT CAUSED THE BIG BANG?

Well, Cheers is long gone, Friends is in syndication, the Andy Griffith Show is a bit dated, Seinfeld is also in syndication, nerds are suddenly cool (Zuckerberg, Gates, Jobs, Musk), so a Cheers-Friends-Griffith-Seinfeld-type show about nerdy friends seemed like a good idea. And so it was.

Oh, you mean the Big Bang Big Bang? Why didn't you say so? Either it was a random quantum fluctuation (which needed an observer) or God (who might have been that observer.) So really, you gotta go with God on that one. Otherwise, you're just making shit up.

IS IT ACCURATE TO SAY THAT SOMEWHERE ON EARTH THERE IS ALWAYS A SUNRISE AND A SUNSET OCCURRING EVERY MIN OF THE DAY?

No. At each pole for several months each year, it is either sunny or dark all the time. Sorry, dudes. Use sunblock.

WHAT ARE BIG BANG AND BIG BOUNCE THEORY? ARE BIG BANG THEORY AND BIG BOUNCE THEORY RELATED TO EACH OTHER?

Only insofar as they talk about the origin of the universe. Big Bang is the only theory for which we have any evidence, and so far the evidence is overwhelming. The universe came out of a hot dense state 13.7B years ago, at which point space and time began, the laws of physics arrived, energy and matter arrived, everything coming from nothing in a tiny fraction of a second.

Big Bounce is a hypothetical work-around to Big Bang, suggesting that this version of the universe is just one of a very large? infinite? series of universes that continually bounce into being after the previous version collapses.

There's no evidence at all for Big Bounce, and with the advent of Dark Energy and the increasing rate of expansion of the universe, it seems certain that the universe will not collapse but will continue to expand until the Big Rip and the Heat Death of the universe.

So here's what you've got—one universe, one Big Bang, singular and unique.

HOW MUCH OF SPACE HAVE WE DISCOVERED?

I recently saw a number or two about the size of the undiscovered part of the universe, the unobservable universe outside of the 93BLY universe we think we have discovered something about.

The first number was $10\textasciicircum10\textasciicircum30$, that is, the unobservable universe is $10\textasciicircum10\textasciicircum30$th times bigger than our part.

The second number, in the same article, was $10\textasciicircum10\textasciicircum10\textasciicircum122$nd times bigger.

To put those numbers into perspective, there are just an estimated 10^{30} microbial cells on Earth, which is far more than the number of stars in the observable universe.

The estimated number of photons ever produced by all the stars in our part of the universe is just 4×10^{84}.

So in terms of a percentage, we've discovered essentially nothing about space, especially since 96% of our observable universe is composed of Dark Matter and Dark Energy, the nature of which we know nothing. We've discovered quite a bit about that remaining 4%, if it makes you feel any better.

And we have discovered that the universe had a starting point where space, time, and the laws of physics all came into existence in a tiny fraction of a second, everything coming from nothing. That's impressive, esp when you consider that eventually, all the evidence for Big Bang will be inaccessible to us as the universe continues to expand.

But that's not an answer to "how much of space." It's an answer to "where did space come from."

And if space, the universe is infinite in size, then mathematically we've discovered exactly nothing. And it'll never get better.

That's depressing.

WHY ISN'T THERE ONLY ONE POSSIBLE OUTCOME IN QUANTUM THEORY? HOW DOES THIS MAKE SENSE? HOW CAN THINGS HAPPEN FOR NO REASON AS ARBITRARY DIFFERENCES IN IDENTICAL CIRCUMSTANCES ON THE FINEST LEVEL OF ENERGY AND MATTER?

As Neil deGrasse Tyson said, the universe is under no obligation to make sense to you. You have been taught that humans can figure out everything eventually, and that everything will make sense to us. Turns out that's a naive assumption with little evidence to support it. It's Newtonian and now archaic. The Newtonian universe we with which we interact with our senses is only a tiny slice of the relativistic quantum universe that we actually live in. Nobody likes it, but that's a sign of our naive arrogance. It makes no sense. But that's our fault, not that of the universe.

IF YOU WERE ABOUT TO BE EXECUTED BY FIRING SQUAD, WHAT ONE LAST THING WOULD YOU ASK FOR?

Blanks.

IF YOU WERE STUCK IN A BLACK HOLE AND YOU TURN ON A FLASHLIGHT, WOULD YOU BE ABLE TO SEE THE FLASHLIGHT?

If you could go into a BH (which seems unlikely), then from Cornell:

"Inside the event horizon, time and space change places.

"...we see that everyone inside the event horizon is a psychic.

"This happens because light can travel to you from events in the future, so you can quite literally see them.

"You can't see anything closer to the center than you are because light can't travel away from the center.

"If you look away from the center, though, you see two images of everything—one from the past and one from the future."

WHAT IS THE MOST SCIENTIFIC EXPLANATION FOR THE CREATION OF THE UNIVERSE?

Big Bang, and we don't know. We're pretty good back to about Planck Time, at least as far as theory goes, but before that, don't know, not gonna know. God. God is a pretty good scientific explanation for the creation of the universe. Even if it was a quantum fluctuation, gotta have an observer outside of space, time and the laws of physics. As far as I can tell, only God fits that description. Works even if you've got a multiverse, multiple Big Bangs, a cyclic universe, any and all of the other hypotheses (for which there is no evidence), none of them preclude God and everything makes more sense with God. An intelligent, sentient, self-aware, observing, ordering being outside of space and time. Even the Flying Spaghetti Monster doesn't fit that bill.

IS IT POSSIBLE THAT OUR UNIVERSE IS AN EMERGENT PROPERTY OF QUANTUM MECHANICS?

It has been thought that Big Bang was a quantum event, a random quantum fluctuation. This assumes, of course, that quantum mechanics was in operation at the Big Bang Singularity, outside of space, time and the four fundamental laws of physics, which is a large and unfalsifiable assumption, one that will forever be neither true nor false. Plus, you'd need an observer, and that could only be God. So it's possible, but we'll never know. Unless it's God. Then we'll know.

IF MULTIPLE UNIVERSES EXIST, WHICH ONE ARE WE IN?

They don't. So, we're in the only universe that exists.

It is the exquisitely finely-tuned Big-Bang-produced gotta-have-an-observer universe that is so carefully and, it seems, intentionally programmed to produce and support intelligent life on, apparently, one planet only that the only possible alternative that science seems to be able to come up with is, other universes, and we are in the one that just got lucky.

Actually, in terms of the math, if there are an infinite number of universes, then we are in one that would have inevitably arrived, and there are an infinite number of other universes just like it, along with an infinite number of each of an infinite number of possible variations. So saying that we "got lucky" is not accurate. We only got lucky if we are unique, and that level of luck is not possible mathematically or physically.

So we are either unique (in the correct and strictest sense of the word), or just one of many. Evidence is strongly in favor of the former, and there is and never will be any evidence for the latter.

ARE SCIENTIST ABLE TO MEASURE HOW FAST SPACE IS EXPANDING AND ACCELERATING? CAN IT EVER EXCEED THE SPEED OF LIGHT?

Yes, they can, and yes, it can. Spacetime already expanded trillions of times faster than the SOL during the Inflationary Epoch right after Big Bang. Even now (time being a tricky critter to nail down, so we'll use "now" carefully), the universe beyond the observable universe is expanding away from us at faster than the SOL, so that technically that is also happening in the far reaches of the observable universe. It is gradually disappearing from our view. By this time next week, it'll all be gone, nothing left but the moon.

OK, I kid. But only in the last sentence.

WHAT IS THE MINIMUM SIZE REQUIRED FOR A SINGULARITY (OR A BLACK HOLE) TO MAINTAIN ITSELF AND NOT BE CONSUMED BY ITS OWN HAWKING RADIATION?

I read just recently that evaporating black holes will evaporate all the way down to the singularity before the singularity will itself evaporate.

This means the smallest black holes decay the fastest, and the largest ones live the longest. Doing the math, a solar mass black hole would live for about 10^{67} years before evaporating, but the black hole at the center of our galaxy would live for 10^{20} times as long before decaying. The crazy thing about it all is that right up until the final fraction-of-a-second, the black hole still has an event horizon. Once you form a singularity, you remain a singularity—and you retain an event horizon—right up until the moment your mass goes to zero.

WILL WE FIND ALIEN LIFE?

Just turn on Fox News. All aliens, all the time.

Alternatively, it's most unlikely. Even if there are aliens (very unlikely), the universe is too big for them even to be able to find us, or us them, much less to communicate, and far less to be able to travel here. They don't know we're here, can't find out that we're here, couldn't talk to us if they knew, and can't cross the unimaginably vast reaches of space to get here.

On the other hand, if they did, their biology would most likely make it impossible for them to eat us. We just need one volunteer to be eaten as an experiment to show them how inedible we are.

As though watching Fox News wouldn't convince them.

IS THE INTERPRETATION OF THE CLASSIC BIG BANG WRONG?

No, or at least, not yet. Any theory of science can be overturned or improved with new evidence, but all the evidence we have tells us that BB is the way the universe arrived. Quantum Theory tells us that something can come from nothing, not only at BB but inside our universe (indeed, inside all of the protons and neutrons in the universe) all the time, so everything coming from nothing is not a problem for science, as long as you are OK with the assumption that Quantum Theory was up and humming at the Big Bang Singularity. Which is a large and unfalsifiable assumption.

But the evidence that BB is true is found inside this universe in such a compelling way that there really is no challenger to BB. There are some alternative hypotheses, but there's no evidence that any of them is true. When the CMB (Cosmic Microwave Background Radiation) was discovered and then measured, Stephen Hawking called it "the discovery of the century, if not of all time." Such was its power in validating BB.

So at this point in the history of science, BB may be the truest thing in science.

HOW MUCH SOUND WAS PRODUCED IN THE BIG BANG?

I'm gonna say, none. "Sound" is sound waves propagating through a medium, that is, moving through something that waves can move through like waves moving through the ocean water. In the case of BB, you would need gas clouds, and there were no gas clouds yet, nor even any particles until the first 3 minutes of the existence of the universe.

Plus, if a universe falls in the forest and there's no one there to hear it, does it make a sound?

Black holes make a sound, tho. From Arthur Miller:

"The Chandra Observatory has even 'heard' the sound of a black hole, from the Perseus galaxy cluster, 1500 trillion trillion miles away.

"(The sound) results from explosive activity near the black hole caused by large amounts of gas falling into it from smaller galaxies that it is cannibalizing.

"The sound is … a steady B flat, 57 octaves below middle C.

"The hum of a black hole is over a million billion times deeper than can be heard by the human ear." (from Empire of the Stars, by Arthur Miller)

WHAT IS THE MOST ACCURATE CURRENT MODEL OF THE UNIVERSE VISUALLY REPRESENTED?

It's really hard to represent a 4D universe in 2D. Here's one effort on the left:

It puts our solar system in the center (which is where we see the rest of the universe from).

If you consider the stars on the balloon to be galaxies (or clusters or superclusters of galaxies), then you can picture them moving away from each other as the balloon (the universe) expands. Each one seems to be at the center, and indeed it is, but only its own center, and every star (or whatever) is at its own center.

The trick is that the universe doesn't have an inside, and the balloon does. In this illustration, the 4D universe is just the 3D surface of the balloon, which at a close enough distance will seem flat and only 2D. So ignore the inside of the balloon.

Apart from that, the universe is really only represented accurately using math.

WHAT ARE SOME COOL FACTS ABOUT SPACE THAT MOST PEOPLE DON'T KNOW?

So many. Here's one—space and time can change places, which happens in a black hole. So the singularity around which a black hole forms no longer has a spatial location, only a temporal one. That is, it's not over there. It's tomorrow.

Here's another—"down" is not a spatial direction, it's a temporal direction (mostly). So "down" is the direction in which time passes more slowly, and gravity is basically a time slope more than a space slope.

WILL HUMANS ONE DAY BECOME ADVANCED ENOUGH TO NAVIGATE THE MULTIVERSE?

We may possibly one day become advanced enough to navigate the solar system, that is, to send humans flitting about. It's doubtful that we'll ever make it to our nearest star neighbor. We'll never navigate the Milky Way galaxy, nor do we have any possible dream of making it to any of the 50 or so galaxies in our local group. The observable universe will be forever beyond us, and of course, the rest of the universe, the unobservable part, is not even thinkable.

And since that's just this universe, well, there you go.

Plus, there is no multiverse, and even if there were, there will never be any evidence of even one other universe, much multitudes of them.

So, no. We will never be able to navigate most of what does exist, and infinitely less will we ever be able to navigate something that does not exist.

WHAT WOULD REALITY LOOK LIKE INDEPENDENTLY OF ANY PARTICULAR BEING'S PERCEPTION? IN OTHER WORDS, WHAT DOES THE WORLD REALLY LOOK LIKE, NOT JUST HOW WE SEE IT OR HOW A DOG SEES IT?

Here's a metaphor. Look at the energy spectrum.

Where you live, and what you experience, is in the narrow band of visible light. Everything else is there, but you have no ability to experience it directly. And even then, as others have said, your brain interprets the visible light for you so that it makes sense and you can avoid being eaten by something.

The metaphor of the spectrum works at all levels. It works for sound, obviously.

But it also works for the universe at large. You live in what you'd consider to be a Newtonian universe, a Newtonian construct, where everything makes sense, everything is logical and reasonable.

But the universe at its core is not purely Newtonian—it is relativistic and quantum. But Relativity only is noticeable at very high speeds and in very intense gravitational fields. In your normal existence, it is as though Relativity doesn't exist. But you live in the tiny Newtonian sliver of the Relativity spectrum where most of it is inaccessible to you.

And it is a quantum universe. You think that things are solid and that's what defines existence. And things all look and act solid.

But they're not solid at all. What defines existence is quantum interactions and the fundamental forces; in our case most specifically, the strong and electromagnetic forces, and gravity, of course. There's nothing really here in the universe, no objects, no billiard-ball type particles, no solid things of any type.

There are only interactions, and reality is defined by those interactions. It is the gravitational interaction; gravity is an interaction between spacetime and matter. It is the strong interaction, mediated by the interaction between quarks and gluons. It is the electromagnetic interaction, mediated by photons and electrons. Those are what hold everything together and create the illusion of solidity, the very illusion of reality. Reality, as Einstein said, is but an illusion, albeit a persistent one. The reality you perceive is a persistent illusion, and could you perceive the rest of reality, that illusion would be revealed to you as a cosmos-sized web of interactions.

The universe is interactions. It is relationships, from the very smallest to the very largest. And humans find themselves roughly in the middle of that web of relationships between the very smallest and the very largest. And it seems as though our role as observers cause reality to exist, from the very smallest to the very largest, so that we occupy that sliver, that tiny sliver in the middle, is the most significant of ways.

CAN PANSPERMIA BE ENDORSED BY SCIENCE?

Nothing can be endorsed by science without evidence. So panspermia will be an hypothesis very much like all the other origin-of-life hypotheses, a permanent hypothesis as they will be permanent hypotheses. We are unlikely ever to have a solid theory of the origin of life on Earth, because we are unlikely ever to have compelling evidence. We weren't there. The evidence that might have been found in some sort of fossil record will have been too ephemeral to have lasted until the present day, if it ever existed at all. There's no way for us to know. Even if we orchestrate some convincing experiment, it would still have been orchestrated deliberately rather than happening on its own.

Having said all of that, it is entirely possible for scientists to choose to endorse panspermia and to pretend that it is true. Many are doing it right now with the Multiverse, for which there is no, and likely will never be any evidence. And once upon a time, science did not believe that the universe had a starting point. So anything is possible when it comes to endorsing shaky hypotheses.

DOES ABSOLUTIST THOUGHT REPRESENT THE ARROW OF TIME BUILT INTO THE COMPONENTS WHICH MAKE US HUMAN?

Here are some articles (or bits of articles) on time that will show the interesting complexity of any potential answer based on science. These answers will not help at all.

"It's as if what we did today, changed what we did yesterday. And … the experimental results have spooky implications for time and causality—at least in the microscopic world to which quantum mechanics applies.

"… in the quantum world time runs both backward and forward whereas in the classical world it only runs forward …

"…the measured quantum state somehow incorporates information from the future as well as the past. And that implies that time, notoriously an arrow in the classical world, is a double-headed arrow in the quantum world." (www.sciencedaily.com 10 Feb 2015)

"Entanglement, which seemingly allows instantaneous influences to travel between quantum particles that have previously interacted, defies our intuitive notions of time and space.

"But again, perhaps we have things backwards:

"…that strange defiance of the apparent rules could be because entanglement creates time in the first place." (NewScientist April 2018)

"Time moves as it does because humans are biologically, neurologically, philosophically hardwired to experience it that way.

"It's like a macro-scale version of Schrödinger's cat.

"A faraway corner of the universe might be moving future to past.

"But the moment humans point a telescope in that direction, time conforms to the past-future flow.

"'In his papers on relativity, Einstein showed that time was relative to the observer,' says Lanza.

"'Our paper takes this one step further, arguing that the observer actually creates it.'" (www.wired.com/2016/09/arrow-of-time)

"There is mounting evidence that at the most basic level of reality, time is an illusion.

"Stranger still, laboratory tests … independently point to the idea that time doesn't really exist." (medium.com Aug 2018)

"Einstein defined gravity as the effect of curves in spacetime created by the presence of matter.

"According to the new approach, gravity is an emergent phenomenon.

"Spacetime and the matter within it are treated as a hologram that arises from an underlying network of quantum bits (called "qubits").

"Verlinde traces dark energy to a property of these underlying qubits that supposedly encode the universe.

"On large scales in the hologram, he argues, dark energy interacts with matter in just the right way to create the illusion of dark matter.

"… bendy, curvy spacetime and everything in it is a geometric representation of pure quantum information.

"The bottom line is that 'somehow, you can emerge time from timeless degrees of freedom using entanglement.'" (The Atlantic, Wired, and Quanta, Dec 2016.)

"'Our inability to figure out the correct version of quantum mechanics is embarrassing,' (Sean Carroll) says. 'And our current way of thinking about quantum mechanics is simply a complete failure when you try to think about cosmology or the whole universe.

"'We don't even know what time is.'

"Carroll favours a bottom-up explanation in which time emerges from small-scale quantum interactions." The Guardian, 4 Nov 2015

"…it is necessary to include the properties of the observer, and in particular, the way we process and remember information.

"Our new paper suggests that the emergence of the arrow of time is related to the ability of observers to preserve information about experienced events.

"For years physicists have known that Newton's laws, Einstein's equations, and even those of the quantum theory, are all time-symmetrical. Time plays absolutely no role. There is no forward movement of time. Thus, many scientists question whether time even exists." (Discover Mag Sept 2016)

"According to quantum mechanics, if we have a very precise clock its energy uncertainty is very large.

"Due to general relativity, the larger its energy uncertainty the larger the uncertainty in the flow of time in the clock's neighbourhood.

"Putting the pieces together, the researchers showed that clocks placed next to one another necessarily disturb each other, resulting eventually in a blurred' flow of time.

"This limitation in our ability to measure time is universal…" (sciencebulletin.org 2017)

"Bizarre quantum bonds connect distinct moments in time, suggesting that quantum links—not spacetime—constitute the fundamental structure of the universe.

"Not only can two events be correlated, linking the earlier one to the later one, but two events can become correlated such that it becomes impossible to say which is earlier and which is later. Each of these events is the cause of the other, as if each were the first to occur.

."..quantum correlations are more fundamental than spacetime, and spacetime itself is somehow built up from correlations among events…" (Quanta Magazine, Jan 2016)

"Suppose you want to travel from Helsinki to New York and you have to change your flight in London.

"Normally you would first fly on a plane from Helsinki to London, then wait for some time in the airport in London, then board the flight London-New York.

"But in the quantum world, you would be better off boarding a plane from Helsinki to London sometime after the flight London-New York took off.

"You will not spend any time in London and you will arrive in New York right at the time when the plane from Helsinki lands in London.

"This is mind-boggling but the experiment shows that it is indeed happening." (ScienceDaily.com Feb 2016)

"In regions of strong curvature, time might turn into space. This could happen, for example, inside of black holes or at the big bang. In such a case, what we now know as a spacetime with three dimensions of space and one dimension of time might transform into a four-dimensional 'Euclidean' space." (medium.com Nov 2017)

WHAT DO YOU IMAGINE OTHER LIFE FORMS IN THE UNIVERSE ARE LIKE?

All you have to do is look around at life on Earth even to begin to imagine what it might be like elsewhere. Intelligent life on Earth; dolphins, whales, seals, chimps, bonobos, gorillas, cephalopods (octopus, cuttlefish squid), and then systems of apparently simple organisms wherein intelligent behavior spontaneously emerges, like bacteria, ants, termites, even the cell exhibits self-organization and intentional behaviors. Schooling behavior in birds, fish, butterflies shows emergent intelligence.

Actually, I don't think there is life elsewhere in the universe, and even if there is, they'll never find us and vice versa—the universe is just too danged big.

But there is so much intelligence on Earth in so many varied forms that to restrict it to beings that look vaguely like us shows a deep lack of imagination.

WHICH IS SCARIER, THE FACT THAT WE MAY BE ALONE IN SPACE OR THAT WE MIGHT NOT BE?

Imagine that we manage to destroy everything, and you are all that's left. At least, as far as you can tell. Alone in North America or Europe or Africa. No one else. At least as far as you can tell.

That's where we are. As far as we will ever know, we are all there is. The universe is too vast, the distances too enormous, the number of galaxies too immense, not even to mention the number of planets, for us ever to be able to find any other form of life, or for them to find us. Where do you look, how can you possibly look when even to send a signal out into random space is to be harshly restricted by the speed of light. The furthest any signal from us has possibly reached by now is maybe 100 light years, the number of stars inside of that 100LY bubble miniscule. The Milky Way is 100,000 LYs across, 1000 LYs thick.

And our signal has to reach some alien life form that 1) has the ability to recognize it as a signal from an intelligent race, and 2) has the ability to respond. We've only had this ability for maybe 100 years. Time is too immense for us to hope that we hit a planet with a signal at the exact time that that alien race has also achieved the ability to match ours. They could be too young, too old, they might have destroyed themselves.

That's you, along on Earth, trying to figure out if there's anyone else left alive out there. How do you find them? The Earth is too big—where do you look, how to you get there, do you have a radio, do they have a radio, do either of you have power, and will either of you live long enough to find the other, if they are there at all?

And will you even like each other, or will you quickly begin to consider them to be a food source? And them, you?

We are effectively, if not absolutely alone in the universe. We'll never know. So we should take care of each other.

FAR AWAY, THE LANTERN IS PRETTY, UP CLOSE, IT'S FULL OF HOLES. WHAT ARE THE HOLES?

The closer you get, the more you magnify, the smaller you reach, the more holes there are, until it's just all holes. There's nothing really there but holes.

And interactions. All that's really there is interactions between things that aren't really there.

And that lets the light shine through.

WHEN SPEAKING ABOUT THE FOUR HORSEMEN OF THE APOCALYPSE, WHAT WOULD HAPPEN IF ONE OF THE HORSEMEN TURNED GOOD?

He gets fired, there's a big search for a new horseman, lots of interviews, Skypes, FaceTimes, reading of resumes, checking with previous employers, a background check, and they hire a new horseman. The old, fired horseman goes on Ellen, writes a book, sells the film rights, makes a ton of money, and turns bad again. The new horseman hasn't worked out, so they fire him and hire the old one back. He and the other three decide to protect themselves from getting fired, so they start a horseman union and strike for better pay and hours. And more torturing. They love torturing.

HOW WAS THE BIG BANG FORMED?

Nobody knows. Nobody will ever know. We've got ideas, conjectures, hypotheses. God? Sure. Random quantum fluctuation? Maybe, but you need an observer or some sort of disruption in the Force, and since there was nothing there to disrupt with, you're back to the observer and maybe God. A bounce from another universe? No evidence for it. One of many Big Bangs? Maybe, but still no evidence. A white hole from a black hole in another universe? I read that yesterday, but no evidence for that, either.

The problem is, the laws of physics arrived inside this universe via Big Bang. Make sure you get that—the laws of physics didn't cause Big Bang, they were caused by Big Bang. So we have no physics, no laws, no science to take us to that initial moment to explain it. Spacetime, the laws of physics, energy, matter, everything came from nothing, or essentially nothing, for no reason.

So some sort of divine mover is a logical answer, though not universally popular.

WHAT DOES IT MEAN FOR US, IF INSIGHT FINDS LIFE ON MARS?

InSight isn't looking for life, so any living thing would have to walk up and pose for the camera, but if we found life on Mars via any source, then

We are Martians. That is, life on Earth came from Mars originally, or

Martians are Earthlings. That is, life on Mars came from Earth originally, or

There's life everywhere. Which doesn't seem at all likely, or

Life on both planets came from the same source, some other life source in the universe Which doesn't seem at all likely.

IF I AM IN A BLACK HOLE, AND I SEE AN OUTSIDE OBSERVER'S TIME START TO SPEED UP, IS THIS PROOF THAT TIME IS NOT A HUMAN CONSTRUCT, OR IMAGINARY, BECAUSE TIME IS DIFFERENT IN TWO PLACES IN SPACE?

You don't need to be outside of a BH to see time passing differently— you just need a really good clock or two on Earth. Time passes differently between the NIST in Boulder (up around 8000 feet) and the Greenwich Observatory (down around sea level) by about 5 microseconds per year.

Whether or not this proves that time is not a human construct is moot. The way that we measure time is a human construct, but since we don't know what time is and time may not actually exist, the question of whether or not we construct it is not a part of the question.

WHY DON'T THE STARS AND PLANETS THAT SURROUND BLACK HOLES WHICH HELP US TO IDENTIFY THE BLACK HOLE GET SUCKED IN BY THE SAID BLACK HOLE?

For the same reason that the Earth doesn't get sucked into the sun, and wouldn't even if the sun became a black hole.

First, black holes don't suck. They create a gravitational well in spacetime, a slope down which things slide. It's actually more of a time slope than a space slope, but that's for another post.

Second, they have a gravitational field equivalent to the mass of the star (or whatever) that they came from. So if you could compress our sun until it became a BH, it would have the same mass as it did before and all of the planets would continue to revolve around it as before. And that's how some aliens somewhere would identify our sun - they would see the 8 (or so) planets revolving around ... nothing .. and would figure out that there must be a BH there.

Third, gravity is very weak and falls off dramatically with distance. BHs only have the gravity of the mass of whatever they came from. Our sun would have to start consuming other stellar bodies to grow significantly, and if a star wandered through our solar system and got close enough to the sun to be consumed, the solar system would have been entirely messed up by that star's passage through it.

WHAT IS THE PROBABILITY THAT UNICORNS EXIST SOMEWHERE IN THE UNIVERSE?

Impossible to say. If life is rare to the point that Earth is the only life-bearing planet (and the odds favor that by a large amount), then the probability of a horn-bearing horse is zero, since there isn't one here. If life is common, then it's impossible to say. Since the universe is not infinite in time, the odds are not definite that a unicorn will have been produced somewhere, and we don't know, and will never know how big the universe is in space. So it's impossible to say.

WHAT EVIDENCE IS THERE TO PROVE THAT THERE WAS NO UNIVERSE BEFORE THE BIG BANG?

First, we can't prove anything in science—there are only varying levels of evidence. So if you want evidence, then you've got to go to the GTR, which implies that universe originated from a hot, dense state that was infinitely compressed, a singularity. If you fold QM into that, then it came from maybe a Planck state, still very small, but not infinitely compressed. There is no evidence from either that there was anything before BB, esp since time and space came into existence because of BB, so there was no "before." There are conjectures about what might have been, but they are less real than unicorns and leprechauns.

WHY ARE PEOPLE SAYING THEY DISPROVED "FLAT EARTH" BY SAYING THEY SEE BOATS GO OVER THE HORIZON? BOATS ALWAYS COME BACK WHEN YOU LOOK THROUGH ANYTHING THAT CAN ZOOM IN.

My guess is that flat Earth folks are mostly just messing with us. They like to see everyone get all upset at how stupid it is to think the Earth is flat and and write paragraph after paragraph showing that they are wrong. Now, the flat Earth folks might actually believe that the Earth is flat, but they're not really aiming to be convinced that it's not. They just like to see round Earth folks sputter and spit. Stupid people like to watch normal people get frustrated. It helps them feel less stupid if they can find someone to laugh at. Stupidity is like that. So hey, flat Earthers! You're right! It's flat! Now go find something else to do.

WHY ARE PEOPLE TAUGHT THAT THE BIG BANG WAS THE BEGINNING OF THE UNIVERSE WHEN IT IS SCIENTIFICALLY PROVEN TO BE NOT TRUE? (WWW.FORBES.COM/SITES/STARTSWITHABANG/2017/09/21/THE-BIG-BANG-WASNT-THE-BEGINNING-AFTER-ALL/)

You need to read the article you cited. Carefully. There are issues concerning BB that science continues to try to resolve; the problem is, we have no science to take us any earlier inside that first second of the existence of the universe because the laws of physics didn't exist yet. So science speculates about what might have been in those tiny tiny fractions of a second.

The speculation that is mentioned in the article is called Inflation, which aims to solve the problems listed in the article: the temperature is the same everywhere in the universe in the CMB at the microwave level, the universe is flat rather than curved and there are no heat relics left over from BB, as there should have been.

Inflation proposes that at about 10^{-33} of a second into BB, the universe increased its expansion rate by an extraordinary amount. Though I have seen different rates expressed, the one I like the best says that the universe expanded from the size of a proton to 250 million light years across in 10^{-35} of a second. That explains the temperature and flatness issue, and apparently the relic issue, though I don't fully understand that one.

Problem is, 1) we thought we had found evidence of Inflation in the CMB, but it turned out not to be the case. So although everyone thinks Inflation is true, or something very much like it, there's no evidence for it, and 2) a strong element of fine-tuning is needed for Inflation to have happened, and fine-tuning makes folks nervous.

And what you need to realize is that when the article says that BB wasn't the beginning, it means that there were things going on earlier in that first second than BB, but there is still a belief that the whole thing started from a very small point, and the universe then came into existence. What the article is not saying or implying is that there was time and space eternally existent prior to BB. It's just saying that we don't know what happened in tiny fractions of seconds after it started before what we call BB kicked off. It's frankly a sensational headline designed to catch your attention.

WHAT IS THE THEORY OF THE COMPLEX UNIVERSE?

If you mean, what is Complexity Theory, then that's a whole 'nothing thang, as they say.

Complexity Theory says that complex systems emerged unpredictably in the universe when any organism or system of organisms reaches a certain number of elements. It's called spontaneous emergence, and it's how problems get solved in the universe, especially in the bio-sphere. It does fly in the face of traditional neo-Darwinian evolution, which mandates evolution via random chance mutations and natural selection. Complexity says that far from being random, systems actually move towards solutions of environmental challenge with intent and specificity, doing so vastly more quickly and efficiently than traditional evolutionary theory would expect or allow for. When you fold epigenetics into the picture, then Lamarkian evolution has returned full force, though in a different way than Lamarck spoke of.

So what Complexity Theory is ultimately saying is that problems aren't solved randomly over long periods of time, but very quickly and deliberately over far shorter periods of time, sometimes between one generation and the next. Again, this is principally in the area of living organisms, though I suspect that one could apply it to the direction the universe has taken since Big Bang without too much difficulty.

But in the biosphere, you can find this self-organization at all levels of life, from the level of the gene, molecule, and cell up all the way through the species to man. Yes, even man self-organizes and shows emergent behavior.

WHAT IS THAT ONE THING THAT IS CAPABLE OF DESTROYING THE WHOLE UNIVERSE?

There are two things (that we know of), and they are called the Two Most Dangerous Numbers in the Universe.

The values of the Higgs Boson and the Cosmological Constant of Dark Energy fame.

Change the value of either one of those, and everything just goes away. And it's possible that they could change.

Further, the estimates that we had of each were so far off that one of them is the biggest mistake in the history of human thought.

We missed the value of the Higgs by 10^{17}. It's that much smaller than we thought it was.

And we missed the value of the Cosmological Constant by 10^{120}. There are 10^{120} elementary particles in the universe. That's how wrong we were. It's 10^{120} times smaller than we thought it was.

"If modern physics is to be believed, we shouldn't be here."

"The meager dose of energy infusing empty space, which at higher levels would rip the cosmos apart, is a trillion trillion trillion trillion trillion trillion trillion trillion trillion trillion times tinier than theory predicts. (10^{120})

"...for this explanation to work, the cosmological constant must have a very specific—and tiny—value. In natural units, the cosmological constant is given by 1 divided by a number made of 1 followed by 123 zeros! Explaining this value is considered one of the greatest challenges faced by theoretical physics today." (Nautilus Issue 53, Oct 2017)

"And the minuscule mass of the Higgs boson, whose relative smallness allows big structures such as galaxies and humans to form, falls roughly 100 quadrillion times short of expectations. (10^{-17})

"Dialing up either of these constants even a little would render the universe unlivable." (www.quantamagazine.org/, Nov 2014)

"'The next few years may tell us whether we'll be able to continue to increase our understanding of nature or whether maybe, for the first time in the history of science, we could be facing questions that we cannot answer,' Harry Cliff, a particle physicist at the European Organisation for Nuclear Research—better known as CERN—said during a recent TED talk in Geneva, Switzerland.

"Equally frightening is the reason for this approaching limit, which Cliff says is: 'Because the laws of physics forbid it.'

"At the core of Cliff's argument are what he calls the two most dangerous numbers in the Universe. These numbers are responsible for all the matter, structure, and life that we witness across the cosmos. And if these two numbers were even slightly different, says Cliff, the universe would be an empty, lifeless place." (ScienceAlert, Jan 2016)

ARE THERE AN INFINITE NUMBER OF MULTIVERSES?

Sure! Why not! They're all imaginary anyway, so rather than just an infinite number of imaginary universes making up one multiverse, let's have an infinite number of imaginary multiverses, each made up of an infinite number of imaginary universes! A multiverse of multiverses! And maybe there's an infinite number of those, too! And of those!

That's why the whole multiverse thing is silly. No compelling evidence for even one single solitary other universe will ever be found, so we're just making crap up. We can't even see the overwhelmingly vast majority of this universe, which might be infinite, but we'll never know that, either. I can't even see the back of my own head!

And besides that, the three theories that suggest other universes might exist aren't true yet, either, and recent work on each suggests that other universes might not even logically follow from those theories, should they prove to be true. BTW, that's String Theory (not true yet), Inflation (no evidence yet), and the Many Worlds Hypothesis of QM (which is just wild speculation).

So, no.

IF ALL LIFE ENDS WITH THE HEAT DEATH OF THE UNIVERSE, WHAT IS THE POINT OF DOING ANYTHING?

All life ends many many trillions of years before the heat death of the universe, so you can ask a better question than that. How about, if the universe is pointlessly indifferent to our existence, and in fact to the existence of anything, and even to its own existence, as Richard Dawkins suggests, then what's the point of doing anything?

Thankfully, Dicky Dawkins has his head up his heiny. He says that since God doesn't exist, there is (to quote) "no design, no purpose, no evil and no good, nothing but pointless indifference."

But the evidence for the existence of God to be found in nature is profound, though many choose to believe otherwise. Big Bang (everything coming from nothing via no science for no apparent reason in virtually no time at all), the fine-tuned universe, the observer problem in QM, all of these point to a universe put together to produce and sustain intelligent life, of which we are likely to be the only examples, though it would be fun to have others out there somewhere.

So the point of doing anything is that, apparently, the universe needs you and me and all of us to be here, and was set up in just exactly the right way to do just that.

Your only alternative to regain the much cherished pointless indifference is the multiverse, for which there will never be any compelling evidence and must be accepted on faith.

If you're gonna start accepting things on faith, and one of them holds the promise of your life having a point, I'd go with that one. But that's just me.

IF GRAVITY HAD BEEN ANY STRONGER OR WEAKER UPON SEPARATING FROM THE THREE

OTHER UNIFIED FORCES IN THE BIG BANG, WHAT WOULD'VE BECOME OF THE UNIVERSE?

If gravity had been one part in 10^{60} stronger, the universe would have collapsed back in on itself and there would have been no time for order and structure to have appeared—no stars, no galaxies, no planets, no elements, no life.

If it had been one par in 10^{60} weaker, the universe would have expanded too much too fast, and all of the hydrogen particles (and helium and bits of other things) would have been too spread out for gravity to draw them together into gas clouds, and hence into stars, elements, planets, galaxies, and life. So, no stars, no galaxies, no planets, no elements, no life.

That's a 1 with 60 zeroes after it. (I've also seen 10^{24}.)

Any tiny change in gravity, and there's nothing in the universe.

There are over 200 other constants of nature, a tiny change in even one of which would have resulted in no life, no order, no structure in the universe.

It's called "fine-tuning," and that's the way the universe is.

IF EXTRATERRESTRIALS ARRIVED AND CLAIMED THEY WERE THE GODS THAT CREATED MANKIND, WOULD YOU BELIEVE THEM?

I'd need to see some ID.

IS REALITY UNKNOWABLE, FOREVER HIDDEN BEHIND THE VEIL OF OUR ASSUMPTIONS, PRECONCEPTIONS, AND DEFINITIONS?

"... experiments led by a group at the University of Vienna, Austria, provide the most compelling evidence yet that there is no objective reality beyond what we observe. Rather than passively observing it, we in fact create reality." (NewScientist 2007)

"It starts with quantum mechanics, which is our best theory of elementary particles and the microscopic world. There's this thing in quantum mechanics that says, before you look at an object it's not in any definite location. It's in a wave that you can think of as a superposition [overlap] of all the different locations it could be in.

"So it might be more likely than you observe it in one place or another, but it's not actually located at any particular place until you observe it.

"It's really weird to think that the behavior of this thing is different depending on whether you're looking at it or not.

"That's the fundamental weirdness of quantum mechanics: that objects behave one way when you're not looking at them, in another way when you are." (Sean Carroll, Discover Mag, Oct 2019)

"(In) April (2015), Nature Physics reported on a set of experiments showing a similar effect using helium atoms.

"Andrew Truscott, the Australian scientist who spearheaded the helium work, noted in Physics Today that '99.999 per cent of physicists would say that the measurement... brings the observable into reality.'

"In other words, human subjectivity is drawing forth the world." (Margaret Wertheim, aeon.com, Dec 2015)

"This raises all sorts of hairy questions. For a start, what counts as a measurement that takes us from probability to certainty?

"Quantum experiments have shown that it seems to involve not just doing something with a measuring instrument, but also consciously noticing the result." (NewScientist Dec 2018)

"Fundamentally, I have an ideal of what a physical theory should be," says Nobel laureate physicist Steven Weinberg. "It should be something that doesn't refer in any specific way to human beings...

"It shouldn't have human beings at the beginning in the laws of nature.

"And yet I don't see any way of formulating quantum mechanics without an interpretive postulate that refers to what happens when people choose to measure one thing or another thing."

But for now, at least, quantum mechanics largely seems to withstand every test.

"No, we're not facing any crisis. That's the problem!" Weinberg says. "In the past, we made progress when existing theories ran into difficulties. There's nothing like that with quantum mechanics. It's not in conflict with observation at all.

It's a problem of failing to satisfy the reactionary philosophical preconceptions of people like me." (Scientific American July 2018)

"For almost a century, physicists have wondered whether the most counterintuitive predictions of quantum mechanics (QM) could actually be true.

"Only in recent years has the technology necessary for answering this question become accessible, enabling a string of experimental results—including startling ones reported in 2007 and 2010, and culminating now with a remarkable test reported in May—that show that key predictions of QM are indeed correct.

"Taken together, these experiments indicate that the everyday world we perceive does not exist until observed, which in turn suggests a primary role for mind in nature.

"Now that the most philosophically controversial predictions of QM have—finally—been experimentally confirmed without remaining loopholes, there are no excuses left for those who want to avoid confronting the implications of QM.

"Lest we continue to live according to a view of reality now known to be false, we must shift the cultural dialogue towards coming to grips with what nature is repeatedly telling us about herself." (Scientific American May 2018)

"Today I think we are beginning to suspect that man is not a tiny cog that doesn't really make much difference to the running of the huge machine, but rather that there is a much more intimate tie between man and the universe than we heretofore suspected... The physical world is in some deep sense tied to the human being. (John Wheeler)

"As I wondered why Penrose keeps hammering away at his theory on consciousness after all these years, I asked him if he thinks there's any inherent meaning in the universe. His answer surprised me. 'Somehow, our consciousness is the reason the universe is here.'" (Roger Penrose)

"Thus we see that without introducing an observer, we have a dead universe, which does not evolve in time...

"We are together, the universe and us. The moment you say that the universe exists without any observers, I cannot make any sense out of that. I cannot imagine a consistent theory of everything that ignores consciousness...

"In the absence of observers, our universe is dead." (Andrei Linde, Stanford.)

IS THERE ANY WAY TO DETERMINE HOW MANY TIMES THE UNIVERSE HAS BEEN CREATED?

Yes. Once. Any other ideas to the contrary are pure speculation without evidence or hope of any.

HOW DO YOU THINK THAT LIFE GOT HERE IN THE UNIVERSE?

Nobody knows, and nobody will ever know for sure. We will have theories, hypotheses, conjectures, wild guesses, speculations, but we will never have sure knowledge. Nobody was there and there's no way to go back to figure it out. The birth of the universe via Big Bang, in contrast, left its imprint all over the universe, detectable, measurable, analyzable. So we can trace the history of the universe back well inside the first second of its existence.

What do I think? The universe has very distinct characteristics that are biocentric. In fact, there's a branch of science called biocentrism, another called the Anthropic Principle (strong version), both based on the physics of the universe. They are all about fine-tuning and observer-dependency, among other things. Without going into details, a quantum universe needs intelligent life within it in order for reality itself to exist. So one thought within the science community is that the universe needs life and was set up very precisely in order for complex life to arrive at some point.

A second very salient point is that life is apparently rare in the universe. There are those who suggest for a variety of reasons that we might be unique. Certainly we have seen no signs of intelligent life. The largest part of that is that the universe is so big that we literally can only have found any intelligent life forms within 100 light years of Earth, which is spectacularly tiny in a galaxy of 200,000 light years diameter and a universe where the closest galaxy to us is 2.5 million light years away, a universe with 2 trillion galaxies that we will never be able to have any meaningful contact with, in fact, a galaxy where we will never have any meaningful contact with most of it.

And that's all the observable universe.

So what I believe is that life is probably rare, we'll never know, but the universe has a necessary compulsion to complex life and is set up to produce it.

As a lecturer on all of this plus Chaos and Complexity Theories, I also tend to think that life is emergent, inevitable, but not inevitably as it has emerged, just intelligent and sentient.

How did it emerge? Complexity, as the Edge of Chaos as it has been called, tells us that if you get enough parts together in one place that something greater than the sum of those parts can emerge.

Life is just particles, but somehow those particles combined eventually together to fall over the edge from chemistry into biology, from dust to life.

I believe this was inevitable because the universe doesn't exist without complex life.

I also believe in a God who choreographed it from the first moment. It was not a design. It was and is an emergent process where the outcome, in a Chaotic and Quantum universe, was undefined but inevitable.

Frankly, that's brilliant.

HOW COULD YOU PROVE THE WORLD WASN'T CREATED YESTERDAY?

Quantum Theory actually allows for this. Yesterday, five minutes ago, right when you started reading this sentence, 6000 years ago, whatever. Brian Greene talks about it in one of his books:

"Either the universe came into existence as a random quantum fluctuation this precise instant completely ordered, with all of history and our memories created out of nothing, or…

"Big Bang produced instantaneously a highly ordered universe that immediately started clumping into universal disorder, but higher forms of local order."

It's an entropy issue. Either the universe arrived after BB with very close to zero entropy, or it arrived much much later with much less entropy, which is actually more likely.

But what we believe is that the universe arrived with almost zero entropy, which is pretty remarkable all by itself.

WHAT ISN'T UNDERSTOOD ABOUT THE UNIVERSE?

Well. Lots.

Dark Energy. Dark Matter. That's 95% of what makes up the universe right there, so we're down to 5%.

Quantum Theory. The incompatibility of quantum theory and relativity.
Life. Where it comes from. How it got it here. What it is.
Energy. Forces. Particles. What are they, really?

Humans. Are we an accident or a mandate?

Why is there a universe? Is there only one? Why Big Bang? Is there only one?

Fine-tuning. The Observer Problem in QM.

Why can we describe the universe using math? What the heck is math and where did it come from?

Why is there something rather than nothing? Why did the universe do anything?

Why do we have laws of physics? Why are the constants of nature so finely tuned?

What is time? Does it exist? Why does it go both ways at the quantum level but only one way for us?

Why can quantum particles do all of that weird stuff, and we can't do any of it, when we are made of quantum particles?

Is there really anything here? What is the nothing?

Donald Trump? Really? He's either a perfect example of evil actually existing in the universe, or a great argument against the existence of a loving God. And you can't have it both ways.

HOW DO I BUILD A UNIVERSE?

Go to Costco. You maybe should take a truck.

DOES ANTIMATTER GET SUCKED INTO BLACK HOLES TOO?

Well, as some have said, black holes don't suck, so, no.

But antimatter does fall down into the gravitational well dug in spacetime by a BH, and that's the source of Hawking Radiation.

Quantum theory tells us that matter and anti-matter particles are constantly being created and destroyed, each pair lasting on the order of 10^{-21} of a second before annihilating back to energy.

When this happens near (very very near) the Event Horizon of a BH, then the anti-matter bit can go onto the EH, and the matter bit stays in the universe. Thus, the BH loses mass and the universe regains lost information, after a fashion. BHs thus evaporate very very slowly (10^{67} years for the one in the center of the Milky Way) via Hawking Radiation.

WHAT ARE SOME EXAMPLES OF NON-NEWTONIAN MECHANICS?

I'm gonna go with Relativity and Quantum Mechanics. Plus, my mechanic's name is not Newton. I think it's Fred. I don't think I've ever had a mechanic named Newton. So all of my mechanics have been non-Newtonian.

IN EVER CHANGING WORLD OF SCIENCE AND PHYSICS WHERE NEW THEORIES COME IN EVERYDAY, HOW CAN WE BE SURE THAT SCIENCE IS THE ULTIMATE TRUTH?

As others have said, science is not about finding truth; it's about finding out how things work and making predictions on that basis. Here's a few quotes from scientists and science magazines that illustrate the nature of science:

Richard Feynman: "What we call scientific knowledge today is a body of statements of varying degrees of certainty. Some of them are most unsure; some are nearly sure. But none is absolutely certain."

Ethan Siegal: "You've heard of our greatest scientific theories: the theory of evolution, the Big Bang theory, the theory of gravity.

"You've also heard of the concept of a proof, and the claims that certain pieces of evidence prove the validities of these theories.

"Fossils, genetic inheritance, and DNA prove the theory of evolution.

"The Hubble expansion of the Universe, the evolution of stars, galaxies, and heavy elements, and the existence of the cosmic microwave background prove the Big Bang theory.

"And falling objects, GPS clocks, planetary motion, and the deflection of starlight prove the theory of gravity.

"Except that's a complete lie.

"While they provide very strong evidence for those theories, they aren't proof.

"In fact, when it comes to science, proving anything is an impossibility.

"In science … you never know when your postulates, rules, or logical steps will suddenly cease to describe the Universe. You never know when your assumptions will suddenly become invalid. And you never know whether the rules you successfully applied for situations A, B, and C will successfully apply for situation D.

"It's a leap of faith to assume that it will, and while these are often good leaps of faith, you cannot prove that these leaps are always valid. If the laws of nature change over time, or behave differently under different conditions, or in different directions or locations, or aren't applicable to the system you're dealing with, your predictions will be wrong. And that's why everything we do in science, no matter how well it gets tested, is always preliminary.

"Nothing in science can ever truly be proven.

"It's always subject to revision."

Sean Carroll: "I would say that 'proof' is the most widely misunderstood concept in all of science.

"The fact that science never really proves anything, but simply creates more and more reliable and comprehensive theories of the world that nevertheless are always subject to update and improvement, is one of the key aspects of why science is so successful."

NewScientist Mag: "Most published scientific research papers are wrong, according to a new analysis … there is less than a 50% chance that the results of any randomly chosen scientific paper are true.

"…small sample sizes, poor study design, researcher bias, and selective reporting and other problems combine to make most research findings false. But even large, well-designed studies are not always right, meaning that scientists and the public have to be wary of reported findings.

"We should accept that most research findings will be refuted."

Aeon.com April 2017: "Science is in the midst of a data crisis. Last year, there were more than 1.2 million new papers published in the biomedical sciences alone, bringing the total number of peer-reviewed biomedical papers to over 26 million.

"However, the average scientist reads only about 250 papers a year. Meanwhile, the quality of the scientific literature has been in decline. Some recent studies found that the majority of biomedical papers were irreproducible.

"The twin challenges of too much quantity and too little quality are rooted in the finite neurological capacity of the human mind. Scientists are deriving hypotheses from a smaller and smaller fraction of our collective knowledge and consequently, more and more, asking the wrong questions, or asking ones that have already been answered.

"Also, human creativity seems to depend increasingly on the stochasticity of previous experiences—particular life events that allow a researcher to notice something others do not."

HOW ARE GRAVITATIONAL WAVES RELATED TO GRAVITATIONAL ATTRACTION?

Gravity is an interaction between spacetime and matter—put something made of matter (i.e. has mass) into the middle of spacetime (the universe), and a gravitational well or a hole is dug around it. Gravity is a time-slope down towards the center of mass. So gravitational attraction is just two things with mass trying to roll down each other's gravitational wells towards the center.

Gravitational waves are generated by the two things moving through spacetime and making ripples, but it takes a really large merger of two things to generate what turn out to be detectable gravitational waves, because even at their very largest, the waves are very very small. When 1.3B years ago, two black holes merged (attracted to each other's gravity, or better still, rolled down each other's gravity wells), the merger generated for a brief moment as much energy as all the stars in the universe, but the waves produced were only 1/1000th the width of a proton.

So attraction produces merger produces waves.

BEFORE THE BIG BANG, WAS THERE A BLACK HOLE TYPE SINGULARITY THAT SIMPLY JUST "BLEW UP" TO CAUSE THE CREATION OF THE UNIVERSE? AND IF SO, DOES THIS MEAN THAT WE LIVE INSIDE A BLACK HOLE THAT EXISTS INSIDE ANOTHER UNIVERSE?

Asking questions like these is getting to be like asking about Bigfoot or Area 51. Whew.

Anyway. The Singularity that gave rise to BB and hence to the universe has no characteristics that we can talk confidently about. Roger Penrose at Oxford back in the '70s came up with Black Hole singularities, and Stephen Hawking realized that BB would have come from a singularity, but where we know the origin of BH singularities, we don't with respect to the BB singularity. We simply can't say. It didn't blow up—it expanded really really rapidly to cause the universe, and it continues to expand.

But our theories about what's inside BHs don't seem to match what we see around us. Inside a BH, time dominates over space, and at the singularity, they change roles so that the S no longer has a spatial location, just a temporal one. So it's not "over there"—it's tomorrow, it's in the future. You can see things as they were and as they will be simultaneously inside a BH, both into the past and into the future. And you very quickly will end up in the future having a quick and intimate moment with the S, which will not turn out well for you.

But I have seen a suggestion that every universe is in a BH, which is in a universe, which is in a BH, and it's BHs all the way down.

I don't think so. But we'll never know.

DO QUANTUM MECHANICS AND THEORY OF RELATIVITY COMPLEMENT EACH OTHER?

Congratulations on spelling "complement" correctly!

QM and R are actually incompatible at the quantum level, which is the only place where they intersect. It's one of the great mysteries of physics, one that String Theory would fix if String Theory turns out to be true, or, better said, falsifiable.

They are incompatible at the level of the very smallest, where R likes a smooth, featureless thing and QM demands quantum foam, a seething field of virtual energy and particles. And QM is right, so R is going to have find the right path.

They are incompatible at the Singularity in Black Holes and at the one out of which Big Bang produced the universe.

So it's a bit of a problem. Fix it and win yourself a nice prize.

HOW LONG WOULD IT TAKE TO CROSS THE MULTIVERSE?

Since we'll never know if there is a multiverse, and since odds are against it, and since we don't know if other universes have time, and since there's no way to know what's between other universes and if this includes time, this question has no possible answer. So, 42. Pick your own unit of time.

IF THE OBSERVABLE UNIVERSE NOW (92 BILLION LIGHT YEARS) WAS SCALED DOWN TO THE SIZE OF A MUSTARD SEED, WHAT ARE YOUR THOUGHTS ON THE TRUE SIZE OF THE ENTIRE UNIVERSE?

I've seen the size of the entire universe estimated at 10^{26} times larger than the observable, at 10^30^30 times larger, at 10^10^10^122 times larger, and infinitely larger. A mustard seed is 1–2 mm. So do the math.

HOW DOES THE DOUBLE SLIT EXPERIMENT AFFECT OPINIONS ABOUT THE UNIVERSE AND ALSO WHAT IS BEING DONE TO FURTHER VERIFY THE RESULTS?

Answering part two of your question first, this is one of the most verified experiments in all of science, precisely because it is so extraordinary and has changed everything about what we understand about the universe. And they keep trying, and it always works out the same way. And it is the most fundamental element in quantum theory, the base point, the place where it all began and where it all ends.

Part one: It is becoming increasingly, if controversially and unpopularly clear that observation determines reality, that there is no reality without observation, and that the observers probably have to be something very much like us, and not just measuring devices. Everything seems to be quantum, dependent upon quantum interactions to exist—space, time, gravity, the laws of physics—everything. They're still working on it, but it's headed in that direction. Even the direction of the flow of time might come from observation—human observation.

So first, it seems that humans, and maybe other bright aliens, are needed for the universe itself to exist, and thus that the extremely tightly wound parameters of nature, the constants of nature, are orchestrated to produce us.

That's not a popular view, as you might imagine. But it's hard to argue against if you actually are aware of the issues.

Second, it seems that the universe is not composed of things, but of interactions. That it is a relational place. Everything is an interaction between, um, things that aren't really there except as by-products of interactions.

Everything that we call real, as Bohr said, is made of things that cannot be regarded as real. So the only real things are interactions.

And observation is just an interaction between observer and observed.

And it all goes back to double-slit.

WHAT SCIENCE IS FOUNDED ON QUANTUM PHYSICS?

As it turns out, all of them. Where QM was once just a curiosity, more or less, since the '20s and esp in the last few years we've discovered that QM is at the fundamental root of everything:

Solar thermonuclear combustion/Nucleosynthesis

Laser technology (that means CD/DVDs)

Nuclear energy (both fission and fusion)

Superconductivity

Semiconductor technology, so anything with a chip in it (cell phones, microwaves, dishwashers, trains, planes, automobiles, computers, etc.)

DNA—Genetic engineering, cloning

Enzymes—biochemical reactions in a cell

Evolution—adaptive mutations

Photosynthesis/some animal respiration/bird migration

Sense of smell (directly), sight and touch (indirectly)

Time, space, and the laws of nature, and even the direction of the flow of time.

Energy, matter.

The universe.

Reality itself.

Everything emerges from quantum interactions. In fact, it may be that quantum interactions are the only real things in the universe.

WHAT FACT OR BIT OF INFORMATION ABOUT OUR UNIVERSE AMAZES YOU EVERY TIME YOU THINK OF IT?

Big Bang—everything from nothing in a tiny fraction of a second, giving rise to

an unbelievably finely-tuned universe with the constants of nature so precise that to change any one of them would eliminate life from the picture, and the observer problem in QM, i.e. that the universe needs intelligent observers in it in order for reality itself to exist, and finally

that nothing really exists except interactions, starting with the most important, quantum interactions.

Sorry. That's four bits. I feel bad.

WHY HAVEN'T WE BEEN ABLE TO DEFINE GRAVITY YET? WHERE ARE WE STUCK AND WHAT IS THE PRIZE TO BE WON FOR THE PERSON THAT WILL SOLVE THIS? COULD THIS BE WHERE EINSTEIN THEORY WILL BE PROVEN WRONG?

Gravity is an interaction between spacetime and matter.

As it turns out, everything in the universe exists as a product of interactions, and science is starting to say that there aren't any actual things in the universe (Bohr—everything we regard as real is made up of things that cannot be regarded as real.), but just interactions. The gravitational interaction is one of the fundamental four—the strong, weak, and electromagnetic interactions being the other three, and quantum theory being nothing more than interactions, at the most fundamental level, interactions between observer and observed.

That's not to say that we understand any of it. Not only do we not understand gravity, we don't understand the other three interactions and we surely do not understand quantum mechanics. We don't understand dark energy and matter, the Higgs Boson, why there is something rather than nothing, where the laws of physics came from, Inflation, why there are laws of physics, where did life come from, why did Big Bang happen, and so on.

It's unlikely that Einstein will be wrong on gravity, but it's not unlikely that he'll be short of the truth about it, as Newton was.

Show that, and you might indeed win a nice prize.

I'VE HEARD THAT WE COULD BE IN A MULTIVERSE, AND THAT ANOTHER UNIVERSE COULD HAVE DIFFERENT LAWS OF PHYSICS. DOES THIS MEAN THAT BEINGS WITH POWERS

FROM TV SHOWS OR MOVIES THAT BREAK PHYSICS IN THIS UNIVERSE COULD EXIST IN ANOTHER?

And that's why the Multiverse is a ridiculous concept.

WHAT ARE YOUR THOUGHTS ABOUT PARTS OF OUR UNIVERSE THAT WE CANNOT YET PERCEIVE OR SENSE?

It all fits. We are ignorant of 95% of what's inside the parts of the universe we can see—Dark Energy and Matter are entirely mysterious to us. We are blocked from investigating what caused Big Bang. The two finest fields of science known to man, relativity and quantum theory, are incompatible with each other. Nobody really understands gravity, the strong, weak or electromagnetic forces, that is, where they come from and why they work. We don't understand why the Higgs and the Cosmological Constant have the values they do, but if they changed, then all life goes away. We don't understand where life came from. The pursuit of knowledge is a humbling thing.

So what's out there? We'll never know. More of the same, I expect. Stars, galaxies, clusters, superclusters. So vast and so far that we will never ever know.

WHAT IS THE MOST MEDIUM-SIZED THING IN THE UNIVERSE?

Apparently humans. We seem to be roughly half-way between the largest of things and the smallest of things, that is, the observable universe and the Planck scale. It's not exact, but it is interestingly close. Planck scale is 10^{-35} m, size of the universe 10^{26} m. If you pick the smallest thing we can kind of measure, the neutrino, it's about 10^{-24} m (if it has a size that can be measured, which QM makes tricky), so comparing the neutrino to the universe gets you 10^{26} m vs. 10^{-24} m, two orders of magnitude difference. Humans sit between 1 and 2 meters, so that's pretty much in the middle.

IS IT POSSIBLE TO TRAVEL IN TIME? IS IT POSSIBLE TO GO INTO THE PAST AS WELL AS INTO THE FUTURE?

Time travel into the future at a rate faster than the normal rate of time passing is possible. You just have to accelerate yourself much closer to the speed of light than we are able to do, or very likely will ever be able to do. The fastest any man-made object will travel (it's getting to that point now) is about 200 km/sec. 10% of the SOL is 30,000 km/sec. So we're quite a bit short. In order to increase our speed, we need new methods of accelerating. There's a proposal to send a large fleet of tiny probes to Alpha Centauri by accelerating them to 20% of the SOL using lasers, but the time dilation is still very small. You've really got to get to 90% and over of the SOL for the effect to be significant. And the probes are very tiny, so you're not gonna hitch a ride.

You could also travel into the future by getting yourself near a black hole, but good luck with that. I've heard that the closest BH to Earth is 1500 light years away. The intense warping of spacetime near a BH will send you into the future at a faster rate, but again, gotta get there.

If you spend 6 months on the ISS, you'll come back .007 seconds younger than you would have been, so you've traveled .007 seconds into the future.

Backwards in time travel seems to be possible only by finding yourself a couple of (still theoretical) worm-holes and yadda yadda yadda, but good luck with that. Worm-holes are a terrifyingly dangerous place to put yourself, anyway, even if they do exist.

At the quantum level, the arrow of time goes in both directions so that the future can influence the past, but that's at the quantum level, and you are very much a macro object, so good luck with that. We've no way to control or influence that process, so you can't manipulate the quantum to bounce around willy nilly in time.

So the answer is, it's possible and at least for very small particles, normative in both directions, but neither you nor anyone else will ever do it in a significant way.

WHAT METHODS DO WE HAVE AVAILABLE TO DETECT ET CIVILIZATIONS IN OUR OWN GALAXY AND WHY HAVEN'T WE DETECTED THEM YET?

Lots of cool, imaginative clever methods, all of which have come up with nada.

First, listening. You know, for radio emissions or nuclear pulses from when the aliens blew themselves up or whatever. There've been one or two detections (one of them called Wow!) that we've been unable to validate or eliminate as alien in origin.

Second, looking for evidence on likely planets, planets in the Goldilocks zones of different stars of the right size and temperature. Kinda of like our sun, maybe a bit bigger or smaller. Kind of like earth, maybe a bit bigger or smaller.

Not too hard to find the right kind of star. Turning out to be really hard to find the right kind of planet. There are some options, but nothing really just right. It's actually not easy to find planets at all, and then to see if they are earth-ish is a real challenge. That's where the cleverness gets extreme. But you didn't ask about that.

Looks very much like if you're gonna have life, you gotta have water and carbon. So you gotta find planets that have water and carbon. Not as easy as it sounds.

To find evidence of life, there are other things you can look for. Pollutants is one, like methane in the atmosphere, or CO_2. Artificial lights, like those in large cities, would be another. Elements in the atmosphere that don't appear naturally but as by-products of industrialization would be something to look for. Evidence of a nuclear war, you know, like radiation. There's something called a Dyson Sphere for harvesting energy that we might find—you could look it up. And so on.

So why haven't we found anything?

- There's nothing there. There are no aliens. This seems most likely, based upon the very tight parameters under which life in any form can exist anywhere the universe. The odds are scary low that there's anybody out there but us.
- The distances are too great. There's a tiny blue dot in a picture of a galaxy alleging to be the Milky Way (which it is not - can't take a pic of the galaxy that we're in the middle of) which is how far all of our radio waves have reached since the first moment we started emitting radio waves. So that's the extent of who can hear us, if there's anyone out there. That blue dot is expanding at the speed of light, and it's been doing that since the first radio wave emission by humans in 1906, so, a bit over 110 years. That's

110 light years. It's not very big. Space is really, really big. The Milky Way is 100-200,000 LYs across, 1000 LYs thick. That blue dot is very very tiny.

And we've only had the ability to hear radio waves since about the same time, 1906. So only aliens inside that tiny blue dot that sent us a message would have been heard by us. If they are outside the dot (as all of the rest of space is), then we couldn't have heard them yet. If they are inside the dot but 1) haven't reached the technological level to send and receive radio waves yet or 2) no longer use radio waves to communicate or 3) killed themselves off somehow already, then we've no way to find out if they are actually there.

About 9110 stars. That's all we can see with the naked eye. With telescopes etc. we can see much further, but we are then looking backwards in time; the further we look, the further back in time it is. So we would not be seeing any aliens as they are, but as they were or might have been.

The most distant habitable planet discovered is about 2500 LY away, with another possible at 5000 LY away. (We've discovered, we think, planets in the Andromeda galaxy 2.5 million LY away, but not whether or not they are habitable, and some rogue [without host stars] planets 3.8 billion LY away.) The closest potentially habitable planet is Proxima Centauri b, a bit over 4 LY away.

So all of the potential aliens that we know of are between 4 and 5000 LY away. But the furthest we've reached with radio waves is 113 LY, so the only potential aliens who could have heard from us will have to live within 113 LY. Anybody else, too far away from us for them to have heard from us.

We have been able to hear something if it were there to hear since maybe 1896 (Tesla), so if there had been anything to hear, we've been listening (off and on) since then, but really only deliberately since SETI started in the 1980s.

But anything that we might have heard could have been sent from any time. We're listening backwards in time. By the time we heard a message, the aliens that sent might have been long gone, blown up or climate changed or asteroided or whatever. A message sent from 5000 LYs away originated 5000 years ago. Just think where we were technologically 5000 years ago, and where we might be 5000 years from now.

And still, we've heard nothing. We'd have to match our technology time-line with that of aliens whenever they sent their message out. Not now. Then.

For us to match time-lines is extraordinarily unlikely, unless the universe is littered with life, and if that's the case, we would have heard something.

So either 1) we're alone, 2) we're almost alone, or 3) we're relatively alone, but the distances are too great. And even if we're not alone, the odds of our technology and theirs matching time-lines are very very small, because when distances are that great, time lines don't match.

IS SUGGESTING AN INTELLIGENT DESIGNER AS THE ORIGIN OF THE OBSERVABLE UNIVERSE JUST AS VALID AS SUGGESTING A NATURALISTIC CAUSE LIKE TWO COLLIDING BRANES IN A MULTIVERSE, SINCE THESE ARE BOTH METAPHYSICAL HYPOTHESES?

I very much enjoy watching the mental gymnastics involved in avoiding the question of evidence by some folks.

To wit: When we look for signs of intelligent activity on Earth from pre-history, we look for patterns. When we find paintings on cave walls or tools that show distinct signs of having been painted or fashioned with intent, then we rewrite the history books once again about the ascent of man. Sometimes the paintings or tools show very slight, subtle signs of deliberate manipulation, but we are happy and willing to assume that they were made with intelligent intent.

When we look into space for signs of intelligent life, we look for patterns that would indicate intent—radio broadcasts, Dyson spheres, atmospheric gases that could only come from industry, light signatures, whatever.

And then somehow, when we look at our own universe and find 1) fine-tuning and 2) the need for intelligent observers, we can declare it all to be a big accident, a random blip of no meaning or importance. When Big Bang arrived on the scene of theoretical cosmology, it was rejected because it sounded too much like religion, specifically like the first chapter of Genesis. When we discovered that BB produced space, time, energy, matter and the laws of physics themselves, we did not allow ourselves to wonder if this extraordinary order, these amazing patterns that produced and defined everything had an intelligent originator.

We believe on the one hand that sophisticated patterns come only from intelligence, but on the other that the sophisticated laws of physics could not have come from intelligence.

We are intellectual hypocrites, afraid that we will find God at the root of it all.

So we conjure up other universes, colliding branes and hypothetical batteries of alternative laws of physics out of whole cloth, as though that will solve the problem.

But a God who can make one universe, can make as many universes as he likes. After all, nobody makes just one cookie. Biscuit, for you skeptical Brits.

And this universe is supremely finely-tuned, and requires us to observe it into being.

The response from many is that there can be no evidence of God to be found in nature.

But our universe, and the Big Bang that gave it being, had no natural origin, no laws of science or physics that produced it. Nature didn't exist yet.

And surely the laws of physics themselves, which came into being after BB, because of BB, are fine evidence of patterns produced by an intelligence.

If it's evidence of primitive man painting on cave walls or making a tool out of a rock, then it is evidence of God.

GIVEN WHAT WE KNOW ABOUT "FUNDAMENTAL" PARTICLES (E.G. LEPTONS, GLUONS, QUARKS, ETC.) CAN WE PUT LIMITS ON HOW MANY DIFFERENT CONFIGURATIONS DIFFERENT MULTI UNIVERSES MAY HAVE? CAN WE ASSUME THE SAME FUNDAMENTAL PARTICLES WOULD EXIST IN ALL OF THEM?

First, the multiverse is very unlikely to exist.

Second, even if it does exist, we'll never know—we'll never find any evidence that it does.

Third, the answer to your question depends upon the size of the multiverse. String Theory (which is not true yet itself) predicts 10^{500} other universes, which is very large, but not infinite.

But let's just assume it has infinitely many universes.

One version works exactly like ours. Exactly. It's a duplicate. And there are an infinite number of those exact duplicates.

Other versions are slight different than ours, and others, even more different, and still others, even more different, ad infinitum. And for each one of those, there are an infinite number exactly like it.

And then there are versions where the laws of physics are different. It's likely that they don't work at all.

And then versions with no laws at all, with no time, space, or anything familiar. It's likely that they don't work either. So the only ones that are likely to work have the same laws of physics as ours.

Before you jump all over me to say, hey, what about different laws of physics, you're gonna have to suggest some that would work. Because if you change slightly or eliminate even one of our laws of physics, then either there's no universe at all, or it's just a big empty nothing.

Remember—any variant universe you come up with will have an infinite number just like it.

Infinity is much bigger than you think it is.

IS THE UNIVERSE SHAPED LIKE A TORUS?

There's really no way to know. The observable universe looks like a sphere, but that's because that's all we can detect—its real shape might be something entirely different, or without a definable shape at all. If we have measured it as a torus, that would only reflect what we were able to measure, not what it really looks like.

It's like the Flat Earth folks—they think the Earth is flat because it looks flat from their perspective, and they arrogantly and naively assume that their perspective is the only one possible. But they are small and the Earth is big, and it is not flat.

So any shape that we might think the universe is, is only seen to be so from our tiny perspective. We're not even sure what the Milky Way looks like, since we are trying to see it from inside it. There are bacteria in your colon that will define your overall shape much differently than it really is.

Nobody wants to be defined by colon bacteria.

HOW FAR AWAY IS THE NEAREST BLACK HOLE AND IS IT A THREAT TO OUR SOLAR SYSTEM?

I have read that the nearest one of which we are aware is 1500 LYs from Earth and it is no threat. If our sun magically became a BH (without the supernovae), it would not be a threat to us, either. Its gravitational impact on the Earth would remain the same, because it would have the same mass that it now has.

There are rogue BHs, that is, BHs that are not feeding and so cannot easily be detected. One could sail into the solar system, but we'd notice because the gravity of the system would start to go whacky. That would be a threat.

WHAT IS THE LATEST THINKING ABOUT THE ANTHROPIC PRINCIPLE?

It depends upon whom you ask, and which AP you are talking about.

The Weak AP simply says that the universe has the properties it needs in order for intelligent life to exist within it. There is an implied "hmmmm," but that really leans into

The Strong AP, which says that the universe seems orchestrated to produce intelligent life.

As you might guess, this has generated a certain amount of controversy.

It starts with Fine Tuning, that is, the constants of nature that exist within the universe are remarkably, unbelievably dialed in so that intelligent life might exist. Taking just two, the value of the Higgs and that of the Cosmological Constant, if the values of either were altered by a tiny amount, order, structure and life would go away. Add to that the values of the gravitational constant, the strong, weak, and electromagnetic forces, and a host of others, and the math no longer allows for our universe to be as it is by random accident.

Put the Observer Problem into the middle of all of that, that is, that the quantum universe we live in requires intelligent observers in order for anything to exist, and you arrive at the Strong AP.

Some (Victor Stenger, for example) will say that we humans are just well-adapted to the universe. Others will just say, so what? What's the big deal?

And a host of others will say, well, the only alternative to this spectacularly finely-tuned universe is that there must be many many other universes that are not so fine-tuned; we just got lucky.

There are three hypotheses that allow for other universes/the Multiverse: String Theory, Inflation, and the Many Worlds Hypothesis of Quantum Mechanics.

Sadly, none of these three are true yet, and it is looking increasingly unlikely that they will ever be shown to be true. Though the Multiverse is extremely popular with many very bright physicists, there are many others as bright who consider it to be pseudoscience. It is seen by many as a choice between the Multiverse and the existence of God, though others have pointed out that even if the Multiverse exists, it doesn't mean that God does not.

And it all comes back to Fine Tuning, the Observer Problem, and the Strong AP.

WHY WOULD ENERGY CAUSE THE UNIVERSE TO EXPAND?

We don't know what caused Big Bang, what caused the universe to expand initially from the Singularity to our universe, and we will never know. That era of cosmological history is closed to us—there were no laws of physics yet to have caused it to happen—so we are left with the ultimate Gap; we have no science to explain it.

After the initial super-rapid expansion, the rate of expansion began to slow, as it would have been expected to do, but after 5 or 6 or 7 billion years, the expansion rate increased. We don't know why, but the laws of physics inside our universe tell us that it would take energy in order for this to happen. So we call it Dark Energy. Energy loosely defined is that which causes work to be done, and increasing the expansion rate of the universe is work of the highest order.

Think of it like this—you go down a hill on your bike, and your speed increases because of gravity working on your mass. You reach the bottom and start to coast up the other side of the valley, slowing down because of gravity's drag on your bike. Halfway up, you start to speed up. How did that happen? You started peddling, you put energy into the system.

Somebody starting peddling the universe. Only an input of energy will cause it to happen.

WHAT'S THE LIKELIHOOD OF DISCOVERING LIFE ON ANOTHER PLANET WITHIN THE NEXT 50 YEARS?

The only way we'll find life on another planet inside of 50 years is if we discover it in our own solar system. And that's not likely. Mars is the only real option, and it doesn't look good. None of the other planets are possible life-bearers. If you want to expand your search to moons in the system, then maybe simple life on a moon or two, but then only if one of two things are true:

The universe is littered with life.

Whatever life forms we find in our system come from the same place that we did. That is, that life and we are the same life, and it has spread a bit throughout the system.

Number 1) seems unlikely. We should have seen something somewhere by now if the universe is littered with life.

If there is occasional life in the universe (possible but unlikely), then the odds are vanishingly small that there will be two different life forms in the same solar system.

50 years is far too short a time period for us to find life on other planets. With current technology, it would take 6500 years one way to reach the closest planet. We'll need some orders-of-magnitude better form of propulsion to get anywhere meaningful, and I don't see that in the near future. They're talking about sending little space chips to Proxima Centauri by using lasers to accelerate them to 20% of the speed of light. Even at that amazing speed, it will take 20 years for them to get there, and then they can't slow down—they'll just zip right on by—and it will take 4 years for any message they might be able to send us to get back to us. Imagine zipping by the Earth at 20% of the SOL and 1) finding any evidence of life plus 2) recording that evidence to be able to be sent back home. Maybe, if you got lucky, and if that form of life was technologically advanced enough to have lights on at night and satellites in orbit.

So I'm gonna go with zero.

IF THE MULTIVERSE HYPOTHESIS OF INFINITE UNIVERSES WITH INFINITE POSSIBILITIES IS TRUE, WOULDN'T THERE BE A UNIVERSE THAT WOULD OPENLY CONTACT THIS UNIVERSE'S EARTH?

Asked and answered—This is from Scientific American:

"Let's suppose that the most liberal of multiverse ideas are true. In this case, 'Where is everyone?'"

"Except this time the question is not about where the interstellar travelers are, or why galactic civilizations haven't been spotted. The new puzzle is 'Where are all the pan-multiverse travelers and civilizations?'"

"If reality is actually composed of a vast, vast number of realities, and if 'anything' can, does, and must happen, and happen many, many, times, this presumably has to include the possibility of living things skipping between universes willy-nilly. After all, just because physics in our universe makes that look kind of tricky, it doesn't prevent the physics of a huge number of other universes from saying 'sure, go right ahead!'

"Discounting our all-to-human capacity for self-delusion, there is absolutely no hard evidence that we are being, or ever have been, visited by stuff from other realities.

"So what's the answer? Why isn't this happening?

"It could be that traveling between parts of the multiverse is impossible (except it shouldn't be impossible everywhere, almost by definition) or very, very difficult.

"It could be that no entity who reaches a stage where they could hop between universes actually wants to (except, there has to be someone somewhere who does. Again, almost by definition).

"It could be that we're alone, the only form of life in any reality (except, yet again, almost by definition, a multiverse will contain other life).

"It could also be, very simply, that there is no multiverse." (*Does a Multiverse Fermi Paradox Disprove the Multiverse?* Sciam)

One objection to this line of thought that I have seen raised is that with an infinite number of possible universes, we live in one that hasn't been visited.

Infinity is very big, and fairly confusing.

IS STRING THEORY NOT REFUTABLE AND THEREFORE NOT SCIENCE AS IS ARGUED IN "NOT EVEN WRONG?"?

At the moment, yes, it is not yet science. It's beautiful and solves problems that desperately need solving in physics, but in 40+ years of work, there is not a single String Theory fact. The largest part of the problem is that if strings exist, they are far too small to detect. This is from Nova:

The strings of string theory are unimaginably small. Your average string, if it exists, is about 10^{-33} centimeters long. That's a millionth of a billionth of a billionth of a billionth of a centimeter. If an atom were magnified to the size of the solar system, a string would be the size of a tree. (Oct 28, 2003)

IF ALL THE UNIVERSE WAS CREATED AT THE BIG BANG, HOW ARE SOME ROCKS OLDER THAN OTHERS?

Rocks weren't created at Big Bang. Particles came into being in the first three minutes, but it took nearly 400,000 years for the universe to cool enough for simple elements to form. Then it took billions of years for stars and galaxies to form, and millions and billions of years for stars to produce more complex elements out of which rocks would eventually be made as part of planets. And rocks don't just pop into existence—they come from a variety of different sources, some carbon-based biological forms, some not—and arrive at different times in different places as sediment, run-off, volcanic eruptions, rock concerts, rock music, Rocky & Bullwinkle, all kinds of ways. So although the very elementary particles (hydrogen, helium, traces of other stuff) can be traced back to BB, the more complicated particles came from different stars at different times.

CONSIDERING BLACK HOLES TRAP TIME, WHAT WOULD HAPPEN IF I WERE TO ESCAPE A BLACK HOLE?

It doesn't seem to be possible for anything to actually enter a BH—everything ends up painted on the surface of the event horizon as a 2D hologram—but if you were inside, the only way out is to go backwards in time. "Down" in our universe is a time slope far more than a space slope, and down towards the singularity in a BH is a time slope. It doesn't have a spatial location, only a temporal one, and it's always in your future. So the only way to get out is to be able to go backwards in time, which means faster than light. Not possible.

If, however, you were able to get out, you'd arrive at some random moment in time. I suppose if you were able to get out at the same rate you went in, you might arrive at the same moment and meet yourself. More likely that since time dilation gets more and more severe as you approach the EH, you go forward in time, so you'd arrive in the future, probably the far distant future.

But since you can't get in and you can't get out, all you can do is get closer and closer, and thus travel further and further into the future. You'd disappear into the future. And you'd get spaghettified by the tidal forces, which would be ugly for you, but kinda cool to watch, except we couldn't watch, because spaghettification happens in time rather than space; into the future you'd be stretched and shredded.

WHEN A STAR EXPLODES IN A SUPERNOVA, WHAT BECOMES OF THE STAR? DOES IT BECOME A PULSAR, A BLACK HOLE, A NEUTRON STAR, ETC.?

There are a number of possible fates of stars, depending upon their starting mass:

Our sun will not go supernovae, but will become a red giant, and then a White Dwarf—1 Million times smaller, 60% of original star's mass—very dense

Our Sun + 40% = Blue Carbon Dwarf—10^{34} carats—1 diamond = the Himalayas

Our Sun + 44% = Neutron Star—Bose-Einstein Condensate—1 matchbox = 3 trillion tons

Quark star—never seen
Strange (quark) star—theoretical (as are the next 4)
Boson Star—Bose-Einstein Condensate
Gravastar—Gravitational Condensate Vacuum Star
MECO: Magnetospheric Eternally Collapsing Object
Planck Star
And finally, a Black Hole.

HOW DID ALL THE MATTER IN THE UNIVERSE COME FROM NOTHING?

It's just an expression of standard hot Big Bang cosmology, which is widely known and rarely understood.

Here's what BB says—BB produced space, time, the laws of physics, energy and matter.

That's, well, everything. Apart from that, there isn't anything. Hawking said that there was nothing at the BB Singularity. Folks may want to wish that away because it violates their Newtonian, Copernican sensibilities, but wishes do not redefine reality.

As it happens, something comes from nothing all the time in our universe. It's just a quantum thing. Here's some interesting data points on that:

(WAPO)—Cosmic Question Will the Universe Go Crunch, or Will 'Dark Energy' Rip It Apart? Either Way, Your Head Will Pop

"The ultimate question is why the universe exists at all. (Alan) Guth says the equations for inflation allow the universe to expand forever, … but the equations don't permit an eternal past. There has to be a beginning, back there somewhere. Where did that first pulse, that first little spark, come from?

"Guth says he's worked out scenarios in which the laws of physics allow something to pop into existence from nothing. But he adds, 'I was implicitly assuming that the laws of physics already existed, even when there was no space, no time, no matter. I think that's all I need to assume. Nevertheless I am clearly making a big assumption there. That does raise the question of what caused the laws of physics, where they came from.'

"What's his answer? 'I don't have the foggiest idea.'

"So let's quickly summarize where we are: We're not sure if the universe will keep expanding or start contracting. We don't know if The End, if there ever is such a thing, is many billions of years away or many trillions of years away. We don't know if constants are constant. We don't know why there are laws of physics. We don't know why there's something rather than nothing.

"And thus we might guess that scientists will not soon put theologians out of business."

Physicist Lawrence Krauss released a work titled *A Universe From Nothing* in which he sought to explain that it's not only possible but likely, that our Universe burst into existence from absolute nothingness—not even three-dimensional space existed before our Universe was born.

(Popular Mech 10/2017) "…the uncertainty principle works for other quantities, too. The same principle applies to energy and time. The more you know about a particle's energy, the less you know about when it is, and vice versa. Here, something weird happens: If you know that there will never be a particle at a particular point, suddenly that point could have any amount of energy, sometimes enough to create a particle anyway.

"These particles are called 'virtual particles,' and they're basically quantum fluctuations. Once you make enough 'nothing,' the universe starts trying to find a way to fill it, even if that means creating particles out of thin air to do it.

"There's also the possibility that the entire universe is just one big virtual particle."

(Scientific American mag): "An extreme case of particles' being unpinpointable is the vacuum, which has paradoxical properties in quantum field theory.

"Look closely at any finite region of an overall vacuum—by definition, a zero-particle state—and you may observe something very different from a vacuum.

"In other words, your house can be totally empty even though you find particles all over the place.

"…the theory predicts a particularly mind-boggling behavior of the vacuum: the average value of the number of particles is zero, yet the vacuum seethes with activity"

(www.zmescience.com) Aug 2019 "… an observer … who is accelerating would observe photons and other particles in a seemingly empty space while another person who is (not moving) would see a vacuum in that same area."

(Discover Mag) "The laws insist that the fundamental constituents of reality, such as protons, electrons, and other subatomic particles, are not hard and indivisible.

"They behave like both waves and particles. They can appear out of nothing—a pure void—and disappear again.

"In truth, the following will tell us that what we think is something is actually nothing, acting like something:

"The central lesson of quantum physics is clear: There are no public objects sitting out there in some preexisting space. As the physicist John Wheeler put it, 'Useful as it is under ordinary circumstances to say that the world exists 'out there' independent of us, that view can no longer be upheld.'" (Quanta Mag April 2016)

We have to learn to think differently. Not only did everything come from nothing via Big Bang, everything that we think is something is actually nothing. Nothing but physical laws, physical interactions.

Welcome to the brave new world of physics. Once again.

IF A CATASTROPHIC EVENT OCCURS IN ANOTHER LIFE I'M LIVING IN A PARALLEL UNIVERSE, HOW WILL THAT AFFECT MY LIFE HERE ON EARTH?

(Note: Similar answer to one on page 22.) Well. There are no parallel universes, and even if there are, we'll never know. So if there are an infinite number of you-duplicates out there, they're already out there suffering joys and catastrophes already, and it hasn't affected your life in the slightest.

Now if you started to go bi-polar because some of the you's are having a much better life than you are, and some of the you's are having a much worse life than you are, then you could get wrapped around that axle and let it mess you right up.

Don't do that.

HOW ABUNDANT IS LIFE IN THE UNIVERSE? MATHEMATICALLY IT'S ALMOST IMPOSSIBLE THAT EARTH IS THE ONLY PLANET SUSTAINING LIFE. WHAT ARE SOME THEORIES AS TO WHY ALIENS HAVEN'T COME?

The popular stories among the masses would disagree with you, but the math and physics of the thing would completely agree.

In order for anything to have happened in the universe at all (something rather than nothing), the universe had to arrive after Big Bang in the lowest state of entropy that it would ever be in, a state where variations in temperature and density varied only by 1 part in 100,000.

The Odds against a low entropic early universe happening by random chance...

...are 10 to the 10 to the 123rd to one, against. That's from Roger Penrose, Rouse Ball Professor of Mathematics at the University of Oxford.

So whence life?

One. It is a finely tuned universe, with over 200 parameters, over 200 constants of nature where if even one of them was off by just a little bit, order, structure, and life go away.

If gravity were different by either 1 part in 10^{24} or 1 part in 10^{60} (I've seen both), then there's either no universe at all, or nothing in the universe. No gas clouds, no stars, no galaxies, no elements, no life.

If the values of the Strong, Weak, EM forces were changed by a tiny fraction, no order or structure in the universe. Change the values of the Higgs Interaction, Lambda (the Dark Energy constant), or Alpha (the fine structure constant), and no life in the universe. And there are something like 200 more constants just like those, where if you change any one of them, no life. We're not talking about biology here—this is all physics. Biology comes much much later in the process.

It is fine-tuned. And why, we might ask? Towards what end?

Two. It is a quantum universe. That means it's an observer-dependent universe, one that needs conscious, intelligent observers in it.

Perhaps we should let science speak to us.

"We must be prepared to take account of the fact that our location in the Universe is necessarily privileged to the extent of being compatible with our existence as observers." (weak version)

"The Universe (and hence the fundamental parameters on which it depends) must be as to admit the creation of observers within it at some stage." (strong version) (The Anthropic Principle from Brandon Carter in 1973.)

"[T]he Anthropic Principle says that the seemingly arbitrary and unrelated constants in physics have one strange thing in common—these are precisely the values you need if you want to have a universe capable of producing life." (Patrick Glynn)

"everything about the universe tends toward humans, toward making life possible and sustaining it" (Hugh Ross)

"A life-giving factor lies at the centre of the whole machinery and design of the world." (John Wheeler)

"What good is a universe without somebody around to look at it?

"If you want an observer around, you need life, and if you want life, you need heavy elements.

"To make heavy elements out of hydrogen, you need thermonuclear combustion.

"To have thermonuclear combustion, you need a time of cooking in a star of several billion years.

"In order to stretch out several billion years in its time dimension, the universe, according to general relativity, must be several billion years across in its space dimensions." (Dr. John Wheeler In Cosmic Search Magazine)

"We could not even imagine a universe that did not somewhere and for some stretch of time contain observers because the very building materials of the universe are these acts of observer-participancy. You wouldn't have the stuff out of which to build the universe otherwise. This participatory principle takes for its foundation the absolutely central point of the quantum:

"No elementary phenomenon is a phenomenon until it is an observed phenomenon." (John Wheeler)

"Ever since Copernicus proposed his revolutionary idea that Earth's place among the stars is nothing special, astronomers have regarded it as fundamental.

"The cosmological principle it has evolved into goes a step further, stating that nowhere in the universe is special.

."...at the moment, the cosmological principle is just that—an assumption.

"There is no concrete evidence that it is true, and the evidence we do have seems increasingly against it." (New Scientist mag, Oct 2015)

"For much of our existence on Earth, we humans thought of ourselves as a pretty big deal. Then along came science and taught us how utterly insignificant we are. We aren't the centre of the universe. We aren't special. We are just a species of ape living on a smallish planet orbiting an unremarkable star in one galaxy among billions in a universe that had been around for 13.8 billion years without us.

"But maybe we were too hasty to write ourselves off. There is a sense in which we are still the centre of the universe.

"Science also teaches us that the laws of physics are ridiculously, almost unbelievably, "fine-tuned" for you and me.

"This extreme anthropic principle posits that the universe is so perfect that it must have been made for us, either by an intelligent creator or, more likely, because of some fundamental feature of the cosmos that drives it towards intelligent life." (New Scientist mag, April 2015)

And so, we have come up with another explanation. Since our universe is so fine-tuned, it must be one of an infinite number of other universes. We are the universe that got lucky. To wit:

"What would you rather believe in, God or the multiverse?

"It sounds like an instance of cosmic apples and oranges, but increasingly we are being told it's a choice we must make. Take the dialogue earlier this year between Richard Dawkins and physicist Steven Weinberg.

"Discussing the fact that the universe appears fine-tuned for our existence, Weinberg told Dawkins: "If you discovered a really impressive fine-tuning... I think you'd really be left with only two explanations: a benevolent designer or a multiverse."

"Weinberg went on to clarify that invoking a benevolent designer does not count as a genuine explanation, but I was intrigued by his either/or scenario. Is that really our only choice? Supernatural creator or parallel worlds? It is according to an article in Discover Mag. "Short of invoking a benevolent creator, many physicists see only one possible explanation," writes journalist Tim Folger.

"Our universe may be but one of perhaps infinitely many universes in an inconceivably vast multiverse."

"Folger quotes cosmologist Bernard Carr: "If you don't want God, you'd better have a multiverse."

"The reason physicists talk about the multiverse as an alternative to God is because it helps explain why the universe is so bio-friendly.

"From the strength of gravity to the mass of a proton, it's as if the universe were designed just for us." (New Scientist mag, issue 2685)

But there's a problem with the whole idea of other universes. There's no evidence for them, nor any conceivable way of coming up with any. To wit:

"the problem with (multiverse) research programs isn't that of direct testability, but that there is no indirect evidence for them, nor any plausible way of getting any.

"(They) have a serious problem on their hands: they appear to be making empty claims and engaging in pseudo-science, with "the multiverse did it" no more of a testable explanation than "the Jolly Green Giant did it." (Scientific American mag)

"There's no way we could ever carry out any experiment to test for the multiverse's existence in the world, because it's not in our world.

"It's an article of faith, and not a very secure one.

"The multiverse is a prop, a way to explain away things that can't otherwise be explained.

"The answers given are all cop-outs; the scientists have decided to keep on living as if the multiverse didn't exist, because if it does exist the implications are horrifying.

"Right now, infinite versions of yourself are dying in really horrible ways.

"What's more likely: a potentially infinite number of useless parallel universes, or one perfectly ordinary God?" (Atlantic Monthly mag, Aug 2016)

"Thus we see that without introducing an observer, we have a dead universe, which does not evolve in time…

"We are together, the universe and us. The moment you say that the universe exists without any observers, I cannot make any sense out of that. I cannot imagine a consistent theory of everything that ignores consciousness…

"In the absence of observers, our universe is dead." (Andre Linde, Stanford, Inflation Theory co-creator)

"I find it quite improbable that such order came out of chaos. There has to be some organizing principle.

"God, to me, ... is the explanation of the miracle of existence, why there is something instead of nothing...

"If God did not exist, science would have to invent Him to explain what it is discovering at its core." (Alan Sandage, Hubble's successor)

"I belong to a group of scientists who do not subscribe to a conventional religion but nevertheless deny that the universe is a purposeless accident.

"Through my scientific work I have come to believe more and more strongly that the physical universe is put together with an ingenuity so astonishing that I cannot accept it merely as a brute fact.

"There must, it seems to me, be a deeper level of explanation. Whether one wishes to call that deeper level 'God' is a matter of taste and definition."

"To postulate an infinity of unseen and unseeable universes just to explain the one we do see seems like a case of excess baggage carried to the extreme.

"It is simpler to postulate one unseen God." (Paul Davies)

Summing up: I don't think life is a miracle. Science tells us that producing life, and ultimately intelligent, self-aware, observing life is what the universe is all about and had been since the beginning.

Is God behind it? Certainly not a few formerly skeptical scientists have lost a large part of their skepticism under the weight of evidence.

No evidence for God, some will say? Big Bang, quantum indeterminacy, complex emergent self-organization, fine tuning, the observer problem—all stand as pretty good evidence. Proof? No, there is no proof of anything in science; just varying levels of evidence.

Some will say that there can be no evidence for God to be found in nature, ever.

Well. There's no evidence of that assumption being true. It's just an assumption.

So feel free, on the basis of science alone, to believe in God as the choreographer of the universe, the master magician who pulled the universe, ordered, structured, aiming towards life, out of a hat like a rabbit, only without the hat, or the rabbit.

WHICH LAWS OF THE UNIVERSE DO YOU BELIEVE IN?

Science is not about belief—it's about evidence. So we accept the laws for which the evidence is strong. Gravity, strong, weak, and electromagnetic forces, quantum theory, and so on.

But we don't accept any of them without reservations, because any of them might be shown to be wrong or, more likely, short of completely true.

We used to believe the universe was infinite in time and space; Einstein's relativity came along, and we now believe (though many would rather not) that the universe had a starting point of time and space—it is not infinite in time (and never will be), and may or may not be infinite in space.

We used to believe (and most still do) that evolution was purely, 100% a product of random mutation and natural selection, and that Lamarck was a poor deluded fool. Now Lamarck is back in the picture, dramatically so, and with the advent of epigenetics and Complexity Theory, many are taking seriously the idea that evolutionary theory is driven more by emergence and spontaneous organization in response to environmental challenge.

We used to believe that the universe was 100% baryonic matter (the normal stuff everything is made of) and empty space. Now we think that baryonic matter is only 4%, maybe 5% of what the universe is made of, that Dark Energy and Matter are 95% or 96% percent of the universe, and we have absolutely no idea what either of them are. And space is not empty—it seethes with virtual energy and particles at the quantum level.

We didn't know what gravity is, but we thought we did. Now we know that gravity is an interaction between matter and spacetime, that spacetime warps in the presence of matter, that the faster you go, the slower time passes, and that time passes more slowly at your feet than at your head.

And we used to believe that everything was made of little bowling-ball-like solid particles that had definite locations and velocities. Now we know that's wrong, that particles can sometimes act like particles, and sometimes act like waves, but it depends upon how you are looking at them, as though they somehow know that you are looking at them. Sometimes they are where you expect them to be, but they could be across the universe, and they could even be both here and across the universe at the same time. And it just gets worse from there.

And we believe that the most fundamental laws of science are quantum, that everything else derives from the quantum, and none of it makes any sense at all.

HOW MUCH OF THE UNIVERSE DO WE STILL NOT KNOW? UNIVERSE AS IN EVERYTHING AND CONCEPTS WE STILL HAVEN'T KNOW OR UNDERSTAND.

Well. If the universe (both the observable and unobservable parts together) is infinite in size, then we know 0% about the universe. That's just math—how much we know (which is something) divided by infinity (which is everything), and that's zero. If it's not infinite in size, it's still very very big, as big as $10^{10^{30}}$ times bigger than the observable universe, and maybe $10^{10^{10^{122}}}$ times bigger. So again, if you divide what we know (which is something) by the rest of it (which is a very big number), you get something so close to zero that it might as well be zero.

If you restrict your question to the observable universe, it's a little better. But just to start out with the sobering bit, between 1933 and 1998, we discovered the need for something we call Dark Matter (1933 and 1975) and Dark Energy (1998). Those total about 95% or 96% or the observable universe, and we know absolutely nothing about either one of them, except that they apparently exist.

So we know something about 4–5% of the observable universe.

Now. We know about Big Bang and the process by which matter and ultimately complex life arrived in the universe, except for how life itself actually arrived. We don't know about that. We know about the constants of nature and the laws of physics, and all of the rest of the sciences that emerge out of that. We know then about (hierarchically) quantum theory, cosmology, physics, physical chemistry, chemistry, biochemistry, biology, geology, and hence into the soft sciences—sociology, anthropology, archeology, psychology. Any of those can and often do change as we learn more.

But. We don't know why Big Bang happened, so we don't know why there is a universe. We don't know why there's anything in the universe, why there is something rather than nothing, why the universe did anything, why the universe is so finely-tuned to produce and support complex life, where the laws of physics came from, why there are laws of physics at all, why the laws of physics work the way they do (that is, gravity and the strong, weak and electromagnetic forces), why there are fields, anything at all about quantum theory or why it is so bizarre and bonkers and still is perfect and describes everything in the universe to perfection and is possibly the source of everything in the universe including possibly the universe itself and OMG you've gotta have an OBSERVER?!?! WTH is that all about? And constants? Where did they come from?

So. The further away on the hierarchy of sciences you get from QM and cosmology and physics, the more we think we know, until you get down to chemistry and biology and geology and we start to get a little cocky and big-headed about how much we think we know, but ultimately…

…we don't know why stars produce elements and hence chemistry, and we don't know where life came from or how it arrived and hence biology, and maybe all we really know anything about is …

… rocks.

Except we can't predict Earthquakes or volcanic eruptions. Which is all rocks in one form or another.

No wonder we like reading about celebrities.

WHAT DO WE KNOW ABOUT THE UNIVERSE SO FAR? PLEASE INCLUDE THINGS LIKE THE BIG BANG, CREATION OF ALL MATTER, EPOCHS, AND FORMATIONS OF GALAXIES.

The universe had a starting point called Big Bang, where everything apparently came from nothing via a Singularity, a infinitely small, dense point with nothing in it that produced space, time, the laws of physics, energy and matter in a tiny tiny fraction of a second. There is no science to tell us why this happened, because there were no laws of physics yet. They came inside the first second.

To reiterate: space and time came into existence at BB.

There is no "before" BB. Since there was no space or time, there was no place or time for the singularity to exist within. It didn't exist in the way that we understand existence.

The universe arrived in an almost perfect state of low entropy, to 1 part in 100,000 in temperature and density variation throughout the entire universe. The universe became cosmos-sized well inside that first second.

The parameters of the universe are insanely finely-tuned to produce order, structure, and eventually intelligent observers.

Quantum Theory tells us the universe needs intelligent observers in order for anything to exist. Nobody likes that very much. There seems to be no way around it.

The universe seems to be purely a product of quantum, relativistic interactions. There's nothing really in the universe; no particles, no solid objects, no objects independent of quantum interactions. There are just "relata," relationships, interactions. Nothing exists except as defined by its relationships with other things, and those things don't exist except as defined by relationships.

DID THE BIG BANG ACTUALLY OCCUR? (ORIGINALLY ASKED AS "WHAT LED TO BB?")

Your question really would be better phrased as "What's your personal, passionate, but essentially meaningless opinion about what led to BB?" Since we don't know and will never know, at best you'll get thoughtful opinions; at worst, rants.

So here's what science will tell us—nothing. We have no idea. It might have been a quantum fluctuation, but science also tells us that you need a disturbance (and there was apparently nothing there to do the random disturbing) or an observation for BB to have happened. In which case, it was either God, the Flying Spaghetti Monster (or "God" for people who don't want the idea of God to be taken seriously), or humans making the observation nearly 13.8B years after BB—that's from John Wheeler. Or QM was not in operation at BB, which is also possible, and perhaps even most likely. In which case, you need neither a disturbance or an observer, but you have even less idea of what might have caused BB.

After that, everything else is wild speculation without evidence.

So God is as good an idea as any. Just be careful what you say after that. "God did it" doesn't mean it was your own version of God.

Now, that's a fun thought.

WHAT MAKES YOU THINK THAT THERE IS AN INTELLIGENT DESIGNER BEHIND THE UNIVERSE?

I'd quibble with the word "designer" since that implies things that can be taken out of context, like something we'd consider to be poorly designed, so either the designer failed that part of the course, or the idea of a designer is absurd.

Years ago, on West Wing towards the end of the series, character Matt Santos was running for prez as a Catholic Latino. He talked about science and faith in a couple of spots. In one, which I can't find on YouTube, someone asked him if he believed in intelligent design. He answered, I believe in God, and I believe he's intelligent.

That's a far better starting point.

SETI is an organization that looks for alien life in the universe. In doing this, they make an assumption—sophisticated patterns are a clear sign of intelligent life. So they look for patterns—radio signals that have complex, lengthy patterns, evidence of Dyson spheres, of artificially produced chemicals in the atmospheres of distant planets, of signs of nuclear war also in the atmospheres, and so on. There are patterns that can look complex but have naturally explainable causes, though we may not immediately be able to figure out what those causes are.

A recent example was the finding of phosphine in Venus' atmosphere. Phosphine dos not occur naturally; it is produced by living processes. So the news rapidly went viral—signs of life in the clouds of Venus? Turns out, there was no phosphine when we looked again and with better tech. But that's a great example.

We do the same on Earth. We find wall-paintings on caves, arrowheads in the dirt, pottery in excavations, and we assume that they were all intelligently produced. We even see what seems to be intelligent patterns of behavior in apes, crows, bacteria, viruses, sea mammals, cephalopods, and we assume a higher level of intelligence must be present than we thought. When we see emergent complex behavior in simple systems, we talk in terms of emergent intelligence, intent, deliberation, even free will.

So we assume that complex patterns have intelligent origins. Why, if the workings of the universe reveal complex patterns, would we not make the same assumption about its origins?

Science has shown us that, first, the universe had an origin that we call Big Bang, and which science rejected because it sounded exactly like something with a divine originator. It then showed us that BB produced, as far as we can tell or will ever know, a unique universe that was low entropy and essentially instantaneous in the time it took for it to become a universe, that produced space, time, the laws of physics, energy and matter and the potential for everything else, and was, and is fine-tuned, observer-dependent, and derived entirely from those physical laws, those natural interactions, a universe that is empty except for the by-products of those interactions. It is a universe with dozens of physical parameters that are fine-tuned to an extraordinary extent, so much so that if even one of them were slightly different, there would be no order, structure, or potential for life.

That sounds intelligent to me. Not all would agree, of course. They would agree that paintings of animals on caves are signs of intelligence. Why they do not see how much more sophisticated the universe is than that, is curious. So they choose to disbelieve in fine-tuning, in observer-dependency, in uniqueness despite the strength of the evidence. Ultimately, one is compelled to wonder if they just really don't want to believe in God as a matter of personal faith.

WHAT DO YOU GUYS THINK ABOUT THE ANTHROPIC PRINCIPLE?

I'm gonna go with this side of the discussion:

Inflation holds that in the instant after the big bang, the universe expanded rapidly—so rapidly that an area of space once a nanometer square ended up more than a quarter-billion light years across in just a trillionth of a trillionth of a trillionth of a second. And maybe much much bigger.

.0000000000000000000000000000000001 sec

2,500,000,000,000,000,000,000 km

(One million trillion trillion trillion trillion trillion trillion times its initial volume)

"To make inflation happen at all requires us to fine-tune the initial conditions of the universe."(Lee Smolin—NewScientist Mag, Jan 2015)

(I've heard several different explanations of the speed of size of the Inflationary expansion. This one is more fun; the others are as impressive mathematically, but not conceptually.)

"Dark matter plays the role of Creator: its gravity is pulling sections of the Universe to buckle back on itself, forming galaxies along the way.

"Dark energy is doing just the opposite. It's fighting the collapse by propelling the universe to expand at an ever-faster rate.

"Luckily for us, dark matter has been winning for most of cosmic time, particularly in the all-important early stages.

"Our Galaxy, the Milky Way, would have never collapsed out of the expanding rush of the Big Bang without the aid of dark matter's pull.

"That means no Sun, no Earth, and no you." (Dark Matter and the Origin of Life—Starts With A Bang!—Medium)

"Here's the problem: The way physicists understand it, the processes that formed those first particles should have produced an equal number of antiparticles, thereby annihilating all matter and effectively canceling everything out.

"But they didn't. That has left physicists scratching their heads for decades trying to ask this most basic question:

"Why does anything exist at all?" (Curiosity.com Nov 2017)

"The reason for our existence—the existence of anything—remains a fascinating unknown even in science." (medium.com, The Missing Antimatter Mystery: Why are we here? Nov 2018.)

"If modern physics is to be believed, we shouldn't be here.

"The meager dose of energy infusing empty space, which at higher levels would rip the cosmos apart, is a trillion trillion trillion trillion trillion trillion trillion trillion trillion times tinier than theory predicts. (10^{123})

"…for this explanation to work, the cosmological constant must have a very specific—and tiny—value. In natural units, the cosmological constant is given by 1 divided by a number made of 1 followed by 123 zeros! Explaining this value is considered one of the greatest challenges faced by theoretical physics today." (Nautilus Issue 53, Oct 2017)

"And the minuscule mass of the Higgs boson, whose relative smallness allows big structures such as galaxies and humans to form, falls roughly 100 quadrillion times short of expectations. (10^{17})

"Dialing up either of these constants even a little would render the universe unlivable." (quantamagazine.com, Nov 2014)

"'The next few years may tell us whether we'll be able to continue to increase our understanding of nature or whether maybe, for the first time in the history of science, we could be facing questions that we cannot answer,' Harry Cliff, a particle physicist at the European Organisation for Nuclear Research—better known as CERN—said during a recent TED talk in Geneva, Switzerland.

"Equally frightening is the reason for this approaching limit, which Cliff says is: 'Because the laws of physics forbid it.'

"At the core of Cliff's argument are what he calls the two most dangerous numbers in the Universe. These numbers are responsible for all the matter, structure, and life that we witness across the cosmos. And if these two numbers were even slightly different, says Cliff, the universe would be an empty, lifeless place." (ScienceAlert, Jan 2016)

"What if the Higgs mass, and by implication the laws of nature, are unnatural?

"Calculations show that if the mass of the Higgs boson were just a few times heavier and everything else stayed the same, protons could no longer assemble into atoms, and there would be no complex structures—no stars or living beings.

"So, what if our universe really is as accidentally fine-tuned as a pencil balanced on its tip…?" (Quanta Mag May 2015)

"The laws of nature are not some sort of Platonic construct, separate and outside of reality. They have emerged as the universe expanded and cooled—and the way that happened depends very much on the precise physical conditions of the big bang.

"Now it happens that the universe we find ourselves in seems to be very delicately fine-tuned in order for complexity and life to emerge.

"This fine-tuning can be traced all the way back to the big bang, nearly 14 billion years ago. So something very special must have happened at that initial moment." (Thomas Hertog, co-author of Stephen Hawking's Last Paper)

"In *A Brief History of Time* (1988), Stephen Hawking … noted that 'if the rate of expansion one second after the Big Bang had been smaller by even one part in a hundred thousand million million, the Universe would have re-collapsed before it ever reached its present size.'

"In short, a change so small it challenges the imagination, and the Big Bang would have turned into a kind of Big Crunch." (Sept 2018)

"Cosmologists have observed that the expansion of the universe is accelerating, and have attributed this to the so-called cosmological constant (CC).

"Quantum physics predicts that the CC should be more than 10^{120} times larger than observed—a value so large it would blow the universe apart before stars or galaxies formed—yet instead it seems to be just right for the formation of giant galaxies such as our Milky Way, and hence for life.

"This led Nobel laureate physicist Steven Weinberg to propose an anthropic explanation for the CC. He argued that we shouldn't be surprised to find ourselves in such an unlikely place, because it is only here that life could exist.

"That explanation has been gaining favour, especially since string theory suggests that our universe could be just one of 10^{500} possible variants, almost all less hospitable.

"'The anthropic principle is the best explanation for this amazing coincidence,' says Alexander Vilenkin, a cosmologist at Tufts University." (Source unknown)

"… the Anthropic Principle says that the seemingly arbitrary and unrelated constants in physics have one strange thing in common—these are precisely the values you need if you want to have a universe capable of producing life."—Patrick Glynn

"everything about the universe tends toward humans, toward making life possible and sustaining it"—Hugh Ross

"A life-giving factor lies at the centre of the whole machinery and design of the world."—John Wheeler

"We could not even imagine a universe that did not somewhere and for some stretch of time contain observers because the very building materials of the universe are these acts of observer-participancy. You wouldn't have the stuff out of which to build the universe otherwise. This participatory principle takes for its foundation the absolutely central point of the quantum:

"No elementary phenomenon is a phenomenon until it is an observed phenomenon." John Wheeler

Does consciousness create reality? The universe may only become real because we're looking at it. (NewScientist mag cover 25 April 2015)

WHAT IS THE PRACTICAL USE OF BLACK HOLES?

Humility.

The General Theory of Relativity predicts a breakdown of known physics at the BH singularity.

We'll never know if that's true. We'll never know what goes on inside a BH. We'll never know what happens at the Event Horizon. We'll never know if BHs actually exist, or if they fuller only almost exist. We'll never know if they evaporate via Hawking radiation. Well never know where they come from or where they go to. We'll never know why all the galaxies seem to have a great big one right smack dab in the middle, or what relationship those BHs have to their host galaxies.

We'll have hypotheses and math and physics theories and all manner of fantastical ideas.

But we'll never know.

Humility. It's not a bad thing.

SOME SAY THAT THE LAWS OF PHYSICS APPEAR TO BE FINELY TUNED TO PERMIT THE EXISTENCE OF INTELLIGENT BEINGS WHO CAN DISCOVER THOSE LAWS—A COINCIDENCE THAT

DEMANDS EXPLANATION. DOES IT REALLY? WHY NOT ACCEPT THIS MYSTERY AS UNSOLVABLE?

It is far too huge of a coincidence to be chance. And it's not trivial like a mud puddle. The chances of a low entropic early universe are $10\wedge10\wedge123$ to one, against. The value of the Higgs is 10^{120} times smaller than predicted. The value of the Dark Energy Cosmological Constant is 10^{17} times smaller than predicted. The odds against the finely tuned universe are $10\wedge10\wedge30$ to one, against.

If the odds of anything are greater than 10^{50}, they're considered to be impossible. The universe is impossible on a number of fronts—low entropy, the Higgs, Dark Energy, matter-antimatter asymmetry, the tightly wound parameters of nature, not to mention the Observer Problem and the Big Bang itself.

As New Scientist mag said, it's either God or the Multiverse. And there's no evidence conceivable for the Multiverse.

So that's an issue science cannot let go.

WOULD THE UNIVERSE STILL EXIST IF NO LIFE EXISTED TO OBSERVE IT?

You'll certainly get a variety of opinions on that question. The obvious, Newtonian, Copernican answer is yes, of course it would, life and the universe have no relation to each other. Both are just accidents of nature and unrelated. And many reputable scientists would agree.

But earlier there were a couple of New Scientist magazine covers with some accompanying text that illustrate to us that Newtonian and Copernican thinking is, well, old school:

Does the universe exist when nobody is looking?

Does consciousness create reality? The universe may only become real because we're looking at it.

Andrei Linde at Stanford sums it up pretty well:

"Thus we see that without introducing an observer, we have a dead universe, which does not evolve in time…We are together, the universe and us. The moment you say that the universe exists without any observers, I cannot make any sense out of that. I cannot imagine a consistent theory of everything that ignores consciousness…In the absence of observers, our universe is dead."

So, apparently not.

The interesting thing is that it's just modern physics (quantum theory) that tells us this is true, and scientists are supposed to accept the implications of their art.

But because we have been trained to be Newtonian, Copernican thinkers, we react with some heat to the suggestion that humans might actually matter to the universe in an existential way, and the debates are intense and sometimes vicious.

This in spite of the fact that we know that Newtonian physics was fundamentally short of describing the universe in its totality, and that this extreme version of the Copernican Principle has been shown to be based not on evidence, but supposition.

And maybe (probably?) because it has once again become possible to talk about a creative intelligence behind it all. Nobody likes that. But that's just bias at work.

WHY DO WE BELIEVE OUR PERCEPTION IS THE ONLY TRUE ONE?

It's the only one that we have. It's the only one we have access to. Even though we know intellectually that everyone else has their own, it mostly impossible to escape from the bubble that your head is in.

The only vague way to get out of the bubble is to take the bubble into places with which you are unfamiliar and maybe even uncomfortable. Travel, live in other countries, learn other languages and cultures. Even in your own town, cross the tracks to see how people who aren't like you live. Get into their shoes, as the saying goes. Expand your horizons.

Intolerance tends to flourish in places of ignorance rather than hatred. When we assume that our way is the only way, then we assume instinctively that all other ways are the wrong way.

DO PHYSICISTS BELIEVE IN FLAT EARTH THEORY?

Well, the universe is flat, and the earth is in the universe, so, no.

IF MATTER IS NEITHER CREATED NOR DESTROYED THEN WHAT THE HELL HAPPENS TO MY SOCKS?

They were virtual socks, a sock and an anti-sock, that emerged from the virtual energy of foot odor. You must have let them touch each other and they annihilated and went back to being foot odor virtual energy. You must wash each sock separately in the future.

HOW DO HUMANS TUNE INTO MEMORY?

Such a great question. Science has pretty much no idea. We thought we had ideas, but they are always too reductive. If I can rephrase your question into, how are memories stored and how then do we access them, well, we have no idea, but it's a better question.

The metaphor that we'd like to use and to be true is to compare the human brain to a computer, but that, again, is far too reductive and doesn't come close to answering the question. Computers store things in binary code, 0s and 1s, but the brain doesn't show any signs of doing anything even vaguely similar.

From Wired magazine about The Brain Atlas project:

"Studying the brain now is like trying to navigate a vast city without any driving instructions. You don't know where you are, and you have no idea how to find what you're looking for.

"Every brain is profoundly unique, a landscape of cells that has never existed before and never will again.

"The same gene that will be highly expressed in some subjects will be completely absent in others.

"This variation is even visible at a gross anatomical level—different people have differently shaped cortices, with different boundaries between anatomical regions.

"If the human atlas is like Google Maps, then every mind is its own city.

"Scientists assumed for decades that most cortical circuits were essentially the same—the brain was supposed to rely on a standard set of microchips, like a typical supercomputer.

"But the atlas has revealed a startling genetic diversity; different slabs of cortex are defined by entirely different sets of genes.

"The supercomputer analogy needs to be permanently retired.

"Scientists are just starting to grapple with the seemingly infinite regress of the brain, in which every new level of detail reveals yet another level.

"'The problem with this data is that it's like grinding up the paint on a Monet canvas and then thinking you understand the painting.

"'You can't help but be intimidated by the complexity of it all. Just when you think you're getting a handle on it, you realize that you haven't even scratched the surface.

"'What you mostly discover is that the mind remains an immense mystery. We don't even know what we don't know.'"

Scientific American, March 2014:

"Tracking how brain cells compute ... is currently an insurmountable obstacle.

"It moves from measuring one neuron to gaining an understanding of how a collection of these cells can engage in complex interactions that give rise to a larger integral whole—an emergent property.

"The brain .... probably exhibits emergent properties that are wholly unintelligible from inspection of single neurons...

"'Your brain does not process information, retrieve knowledge or store memories.

"'In short: your brain is not a computer

"'No matter how hard they try, brain scientists and cognitive psychologists will never find a copy of Beethoven's 5th Symphony in the brain—or copies of words, pictures, grammatical rules or any other kinds of environmental stimuli.

"'The human brain isn't really empty, of course. But it does not contain most of the things people think it does—not even simple things such as 'memories.'

"'Our shoddy thinking about the brain has deep historical roots, but the invention of computers in the 1940s got us especially confused.

"'For more than half a century now, psychologists, linguists, neuroscientists and other experts on human behaviour have been asserting that the human brain works like a computer.

"'The information processing (IP) metaphor of human intelligence now dominates human thinking, both on the street and in the sciences.

"'There is virtually no form of discourse about intelligent human behaviour that proceeds without employing this metaphor.

"'The validity of the IP metaphor in today's world is generally assumed without question.

"'But the IP metaphor is, after all, just another metaphor—a story we tell to make sense of something we don't actually understand.

"'And like all the metaphors that preceded it, it will certainly be cast aside at some point—either replaced by another metaphor or, in the end, replaced by actual knowledge.

"'The faulty logic of the IP metaphor is easy enough to state. It is based on a faulty syllogism—one with two reasonable premises and a faulty conclusion.

"'Reasonable premise #1: all computers are capable of behaving intelligently.

"'Reasonable premise #2: all computers are information processors.

"'Faulty conclusion: all entities that are capable of behaving intelligently are information processors.

"'A wealth of brain studies tells us, in fact, that multiple and sometimes large areas of the brain are often involved in even the most mundane memory tasks.

"'The idea, advanced by several scientists, that specific memories are somehow stored in individual neurons is preposterous; if anything, that assertion just pushes the problem of memory to an even more challenging level:

"'How and where, after all, is the memory stored in the cell?

"'We can begin to build the framework of a metaphor-free theory of intelligent human behavior—one in which the brain isn't completely empty, but is at least empty of the baggage of the IP metaphor.

"'A few cognitive scientists now completely reject the view that the human brain works like a computer.

"'The mainstream view is that we, like computers, make sense of the world by performing computations on mental representations of it, but (they) describe another way of understanding intelligent behaviour—as a direct interaction between organisms and their world.

"'Because neither 'memory banks' nor 'representations' of stimuli exist in the brain, and because all that is required for us to function in the world is for the brain to change in an orderly way as a result of our experiences, there is no reason to believe that any two of us are changed the same way by the same experience.

"'This is inspirational, I suppose, because it means that each of us is truly unique, not just in our genetic makeup, but even in the way our brains change over time. It is also depressing, because it makes the task of the neuroscientist daunting almost beyond imagination.

"'Worse still, even if we had the ability to take a snapshot of all of the brain's 86 billion neurons and then to simulate the state of those neurons in a computer, that vast pattern would mean nothing outside the body of the brain that produced it.

"'...a snapshot of the brain's current state might also be meaningless unless we knew the entire life history of that brain's owner—perhaps even about the social context in which he or she was raised.

"'We are organisms, not computers. Get over it.'" (by Robert Epstein—Aeon Mag)

"Unraveling brain structure and function has come to mean understanding the interrelationship between neurons, rather than understanding the neurons themselves.

"My hunch is that the brain's power will turn out to derive from data processing within the neuron rather than activity between neurons.

"And networks of neurons enhance the effect of those neurons "thinking" between themselves

"...the brain is not a supercomputer in which the neurons are transistors; rather it is as if each individual neuron is itself a computer, and the brain a vast community of microscopic computers.

"But even this model is probably too simplistic since the neuron processes data flexibly and on disparate levels, and is therefore far superior to any digital system.

"If I am right, the human brain may be a trillion times more capable than we imagine...

"... it is time to acknowledge fully that living cells make us what we are, and to abandon reductionist thinking in favour of the study of whole cells.

"Reductionism has us peering ever closer at the fibres in the paper of a musical score, and analysing the printer's ink.

"I want us to experience the symphony." (From Brian J. Ford, biologist, University of Cambridge)

SHOULD I REPORT MY ALTERNATE UNIVERSE DOPPELGÄNGER TO ICE?

Be prepared for your alternate universe doppelgänger to report you to ICE, in turn. Then you both get deported to the other's universe, which seems a bit pointless. So I wouldn't.

IF TIME IS FLUID, IS THERE A PLACE IN THE KNOWN UNIVERSE WHERE TIME DOESN'T EXIST AT ALL? IF SO, DOES MATTER EXIST IN THAT SAME PLACE?

Time doesn't exist for massless particles traveling at the speed of light. Neither does space. For photons etc., all of space and time are the same, uh, place and time. Since the universe is full of photons (about 4×10^{84} of them at last count), then the universe is full of places where time doesn't exist, except that, as noted, for the photons, the universe isn't there, either.

Time is relative to two things—proximity to the speed of light, and to each gravitational field. Time is unique to each point particle, passing differently for everyone. It's only a tiny amount normally, so we don't notice, but there is no absolute time, as there is no absolute motion.

IN AN ALTERNATE UNIVERSE, WHAT WOULD YOUR NAME BE, AND WHY?

I don't know about mine, but yours would clearly be Sargent Kelly, and you would be a badass. (Answer requested by Kelly Sargent)

IF YOUR PAST SELF TIME TRAVELLED TO THE PRESENT, WHAT WOULD THEY TELL YOU?

Dude, you clearly should have taken better care of yourself. And bought Apple and Microsoft.

HOW HEALTHY IS IT TO QUESTION OBJECTIVE REALITY?

There is no objective reality. We each live in a little time bubble and experience everything from our own perspective, interpret everything from that perspective. That includes filters like my own version of everything, my friends, family, community, church/faith/religion, village/town/city, state/province, part of the country, the country itself, the climate and microclimate, historical perspective, schooling, health, historical era, race/ethnicity, economics/wealth/or lack thereof, language(s) we speak, and on and on. We are largely unaware of how influenced we are by the environment in which we live and grew up. It is largely impossible to question that reality, though it is absolutely necessary to be aware of it, and it should be questioned, even though the very process of questioning is going to be filled with unconscious bias and presupposition.

Worse, physics is now telling us that there may be no objective reality whatsoever. Reality, macro-reality, is defined by observations made by sentient, self-aware intelligent beings who have no control over how that reality eventuates. We are being told that there are no actual physical objects in space—not you, not me, not the earth, not anything—but that everything is a result of physical interactions, relationships, starting with the four fundamental forces of nature and continuing into human relationships.

We are bound, contained within that universe, within that 4D universe, and everything we think we understand, we understand through a 4D filter.

So question away, healthy or not, but hold the answers you find at some distance, because they are not true in the way that you think they are.

WHAT ARE THE BIGGEST HOLES IN QUANTUM THEORY?

Yeah, that's the problem. There are no holes. It's perfect. It's always right, it's never wrong, it's the most predictive scientific theory known to man by orders of magnitude, we use it all the time, nearly everything in the universe has slowly revealed itself to be quantum, and everything else is likely to be quantum, and it makes no sense to anyone. Here's a quote from Steven Weinberg that illustrates it perfectly:

"'Fundamentally, I have an ideal of what a physical theory should be,' says Nobel laureate physicist Steven Weinberg. 'It should be something that doesn't refer in any specific way to human beings…

"'It shouldn't have human beings at the beginning in the laws of nature.

"'And yet I don't see any way of formulating quantum mechanics without an interpretive postulate that refers to what happens when people choose to measure one thing or another thing.'

"But for now, at least, quantum mechanics largely seems to withstand every test.

"'No, we're not facing any crisis. That's the problem!' Weinberg says. 'In the past, we made progress when existing theories ran into difficulties. There's nothing like that with quantum mechanics. It's not in conflict with observation at all.

"'It's a problem of failing to satisfy the reactionary philosophical preconceptions of people like me.'" (Scientific American July 2018)

So the only real hole is that nobody likes it because everyone wants it to make some sort of sense.

WAS AN OBSERVER REQUIRED FOR THE BIRTH OF THE UNIVERSE?

Well, one would like to say that it depends upon whom you ask, but lately quantum mechanics has been reinforcing the need for observers in the universe with some finality.

Our addiction to the extreme version of the Copernican Principle makes it very difficult for most folks to acknowledge that reality. The extreme version, btw, says that humans have no special role to play in the universe. This is based on the same sort of flawed thinking that the Church used to say that the earth was the center of the universe. In the latter case, just because God loves you doesn't mean that the earth needs to be the center of anything. In the former case, just because the earth is not the center of the universe doesn't have any information to offer on the importance of humanity in the cosmos.

We are finally figuring out that the extreme version of the CP is just an assumption without evidence, and flies in the face of what QM is telling us.

To wit:

"We could not even imagine a universe that did not somewhere and for some stretch of time contain observers because the very building materials of the universe are these acts of observer-participancy. You wouldn't have the stuff out of which to build the universe otherwise. This participatory principle takes for its foundation the absolutely central point of the quantum:

"No elementary phenomenon is a phenomenon until it is an observed phenomenon." (from John Wheeler)

"Thus we see that without introducing an observer, we have a dead universe, which does not evolve in time…

"We are together, the universe and us. The moment you say that the universe exists without any observers, I cannot make any sense out of that. I cannot imagine a consistent theory of everything that ignores consciousness…

"In the absence of observers, our universe is dead." (from Andrei Linde)

SO, THE UNIVERSE NEEDS OBSERVERS. AS IN YOUR QUESTION, WAS THERE AN OBSERVER NECESSARY FOR THE UNIVERSE TO BEGIN?

Only if QM was in operation at Singularity. That's one answer, or at least a good starting point

A couple of answers to that:

"The ultimate question is why the universe exists at all. (Alan) Guth says the equations for inflation allow the universe to expand forever, ... but the equations don't permit an eternal past. There has to be a beginning, back there somewhere. Where did that first pulse, that first little spark, come from?

"Guth says he's worked out scenarios in which the laws of physics allow something to pop into existence from nothing. But he adds, 'I was implicitly assuming that the laws of physics already existed, even when there was no space, no time, no matter. I think that's all I need to assume. Nevertheless I am clearly making a big assumption there. That does raise the question of what caused the laws of physics, where they came from.'

"What's his answer? 'I don't have the foggiest idea.'" (from WAPO)

Guth assumes that QM was working so that something (the universe) could come from nothing (Singularity). He doesn't touch on the need for an observer.

Here's one who does:

"Nowhere are such questions more acutely unanswerable than at, and soon after, the big bang.

"Collapsing quantum states in the infant cosmos are thought to have played a pivotal part in its subsequent development, determining how stars, galaxies, planets—everything, in fact—eventually formed.

"But how did they collapse with nothing around to measure them?

"'In ordinary quantum mechanics, measurement involves an external device,' says Sudarsky. 'What's playing this role in cosmology?

"'If I don't want to invoke God or something external to the universe, which I don't, I have no place to locate this measuring device.'" (from Daniel Sudarsky)

So Sudarsky says that a measurement, an observation had to occur in order for the universe to arrive, and he is left, reluctantly, with God as the only the real answer.

The other possibility is one favored by Hawking and Wheeler—the universe came into existence via human observation when we arrived on the scene, and the wave function collapsed backwards in time to Singularity, so, 13.8 B years.

Summing up: nothing happens without an observer, not reality, not the universe. So either QM was not yet activated and no observer was needed, humans did it later, with the universe somehow existing in multiple quantum states for 13.8B years until we got here, or God did it, later abrogating the role of observer to us humans, only we have no control over the nature of any reality we observe into being.

Nobody likes any of those options very much, so it's not much discussed or written about.

Of course, the Occam's Razor choice is God. But, well, you know how well that's going to go down.

CAN ANYONE PLEASE EXPLAIN EINSTEIN'S GENERAL THEORY OF RELATIVITY IN LAYMAN'S LANGUAGE?

Special Theory (STR)—the faster you go, the slower time travels, and the more space is compressed.

General Theory (GTR)—gravity is an interaction between space-time and things with mass, things made of matter. (The universe is made of space-time.) So space-time is warped whenever you put anything made of matter into it. The closer you get to the center of gravity (e.g. the center of the earth), the slower time passes. The more mass you have compressed into a small space, the greater the warping of space-time. And time is more important—"Down" is a time slope far more than a space slope.

So for your GPS location finder to work on your smart phone, your phone has to communicate with a satellite that is traveling at about 17,500 mph a couple of hundred miles up. Because the satellite is traveling faster than your phone, its time is going slower. Because it's further away from earth than your phone, its time is going faster. So time is going both slower and faster for the satellite, and both have to be taken into account in order for you to know where you are and where you are trying to go. (Speed wins—on balance, time is going a bit slower for the satellite than for your phone.)

HYPOTHETICALLY, COULD YOU REACH THE SURFACE OF A BLACK HOLE? LET'S SAY YOU WERE IMMUNE TO THE EFFECTS OF THE BLACK HOLE, AND WENT AS CLOSE TO IT AS POSSIBLE, WOULD YOU EVENTUALLY LAND ON THE HIGHLY COMPRESSED MATTER? WHAT COULD IT FEEL/LOOK LIKE?

There are lots of opinions on this. Here's one: a BH is where a large star (there are other options, but let's go with a star) losing its battle with gravity and collapses into a supernova. Much of the matter in the star collapses down into what we call a singularity, which relativity tells us has no size or substance (QM tells us that it has to have some size, albeit very small).

With all of that mass collapsed into infinite density, space-time itself is warped into a sphere around it. The length of the radius of the sphere is dependent upon how much mass is in the singularity.

The surface of the sphere is warped space-time. There's nothing solid there. It's not really a surface. It's called an Event Horizon and signifies the point of no return.

But since we are restricted to traveling on space-time, we must continue to do so as we approach the BH. But space and time are stretched to infinity at the event horizon.

So it would take you an infinite amount of time to get there, and you'd have to cross an infinite amount of space to do so.

So it seems that you can't reach the EH, can't enter the BH. It might be that point particles can enter, but anything with 3-dimensions can't. Maybe. You would be stretched ("spaghettified") into the future at the EH, along with everything else that was being drawn towards the BH.

If you could go inside (which apparently you can't, but what the heck), the singularity is not at the center of the BH, nor any where inside. Space and time have changed places, so the singularity no longer has a spatial location, just a temporal one. It's not in a place—it's in the future. How far in the future? Maybe infinitely far. So you'd never get there. Maybe.

And we'll never know, because we can't go inside, and even if we could, we can't get a message back outside, because for anything to escape a BH, it has to go backwards in time. Which is faster than light.

It will forever be a mystery.

WHAT DOES PHYSICIST KATE SHAW MEAN BY REALITY IS AN ILLUSION?

Here's what she said in NewScientist:

"We never really touch anything. The atoms of our fingers exchange particles with the atoms of what we touch and we experience a force. Our whole interaction with the outside world is an illusion. Everything we experience is pictures made up by and inside our brains, using information from electrical signals from our totally numb bodies."

What you and I consider to be real, is not really real. As Bohr said, "everything we call real is made of things that cannot be regarded as real." We would consider everything to be made of particles, but particles are just temporary ripples in an energy field—they're not really there.

What is there? The forces of nature, better expressed as the interactions of nature. Particles exist as interactions, and since everything is made of particles, everything exists as interactions. It's all gravity, strong and weak forces, electromagnetic forces, the Higgs interaction. There's nothing in the universe but interactions.

So what we think the universe is, is not. We're wrong. It's a convenient illusion, but it is an illusion. We are just layers of quantum and relativistic interactions.

And chocolate. Don't forget chocolate. Dark chocolate. Which fits. Dark Energy. Dark Matter. Dark Chocolate.

IS IT POSSIBLE TO GO INTO THE FUTURE?

Your head travels into the future faster than your feet. All the time. Unless your head is lower than your feet. The Earth's surface is 2.5 years further into the future than its center. For the sun's surface, it's 40,000 years.

So go really fast, much faster than is possible, or go someplace with a LOT more mass than the Earth. Like a neutron star or a black hole.

You can't do either one, but it's theoretically possible.

WHAT IS THE BEST ARGUMENT AGAINST THE EXISTENCE OF FREE WILL?

The question of free will is a brain question, and we are far from understanding the brain. So we are far from knowing whether or not free will exists, and I'm gonna guess that we will never fully know.

The latest studies seem to indicate that we have free will, but there will be more studies that say the opposite, and then more studies that validate it, and so on.

Pierre Simon Laplace took Newton's physics back in the 1700s, Newton's mechanistic, deterministic universe where everything happened because something caused it to happen. So everything was presumed to be predictable, everything ultimately caused by the eternal extant laws of physics in a universe thought to be infinite in time and space. And so Laplace said we could not have free will. At all. Ever. Everything was cause&effect, everything caused by the laws of physics.

Note: This was dependent on the universe being infinitely large and old, especially the old part, with the laws of physics always here, always working. Laplace also got rid of the need for God to have a role to play, since physics did everything.

Then along comes Einstein, Big Bang, and Quantum Theory.

Big Bang told us that universe was not infinite in time, that it had a starting point, and what was worse, the laws of physics actually came into being because of BB. They weren't eternal.

And Quantum Theory told us that the universe was not purely a cause&effect place, that things happen all the time without a cause, and that everything is quantum.

So the potential for both free will and God to exist re-enter the picture. The potential. Undefined, as yet. It is a quantum relativistic universe that is neither infinitely old nor purely cause and effect where the laws of physics have not always been at play.

The question then becomes, do we actually have free will? We have the potential to have it at some level, but do we actually have it?

No one knows.

The vast majority of the choices we make are not free will choices. We are influenced by family, friends, culture, sociology, history, religion (or lack of), magazines, movies, TV shows, the internet and social media, language, the weather, biochemistry, physics, the era we live in, the country, state, county, city, village, and neighborhood we live in, the schools we attend, the workplace environment, culture and sociology, the food and chemicals we ingest, the diseases we have (physical and mental), and so on.

So the question is, it seems to me, is whether or not we occasionally have the possibility to make a free will decision in the face of all the influences we encounter.

You could easily argue, no. We do not. That's a good argument. It might be knockdown.

I will counter that by saying that I believe that deep within us, because it is a quantum, relativistic universe where God is likely to exist, there is the possibility of every now and then making a free will choice, and I would further argue that there are maybe only two free will choices we can make: to do good or evil, or to believe or disbelieve in God. And I would project that maybe all the rest of our choices will be largely determined by those two.

There are those who will say that those choices are made for us by all of the same influences that impact the rest of our choices, and I will concede that is surely possible.

But I will choose to believe otherwise.

BTW, if you or anyone does not believe in free will, then you cannot be a critic of people who do evil things or of people who practice a religion. You cannot be a critic of anyone, if there is no free will.

Nobody lives like that. We all live as though there is free will. Societies function as though there is free will. There is no other way to function. Societies act as though evil exists, and good, and that choices are made that are either good or evil.

So societies act as though God and free will exist, even if the people within them do not believe. And people act as though God and free will exist, even as they disbelieve.

WHAT ARE YOUR THOUGHTS AND IDEAS ON THE BENDING OR FOLDING OF SPACE?

"Space moderators have hidden your answer."

(I didn't know that space had moderators. I'm a bit concerned.)

GOD PARTS

HAS ANYTHING EVER HAPPENED TO YOU WHICH CONFIRMED OR DENIED YOUR FAITH?

Faith is a constant walk between confirmation and denial. Apparent miracles confirm, the lack of miracles deny. Any person who lives at either extreme - never doubting, never accepting - is almost willfully, fearfully blind to those poles, those extremes, those dramatic tensions.

I fear the person who accepts their faith without ever questioning or wondering, because sometimes (many times?), that person becomes the enforcer of his/her belief system on those who do not believe in exactly the same way.

Likewise, I fear the person who denies faith and God, for much the same reason.

I fear the arrogance of both, the judgement, the air of superiority and condescension.

There is so much that we don't know about science, nature, the universe, and so much that we will never know.

There is so much that we do not and cannot understand about God and our faith, so much that we will never know.

But the unimaginable suffering that we see around us in the world challenges my faith.

But the only way that suffering has any hope for having any meaning is if God exists.

And so we swing, wildly at times, between those two poles.

DO YOU BELIEVE IN THE IMPLIED MEANING OF THE SAYING, "GOD WILL NOT GIVE YOU MORE THAN YOU CAN HANDLE"?

It's not Biblical, any more than "God helps those who help themselves." I suppose you could draw it out from a verse or two by isolating them from their context, but that's always a bad idea.

In terms of the math and physics of the thing, the only ones who can validate it as true are those who have survived something horrendous and can testify about it. Those who did not survive are not here to offer a counter-opinion. So it's a biased group. You wanna have a conversation with those who did not survive the Holocaust? The killing fields of Cambodia? The slaughter of natives in the US and around the world? Slavery? The Black Plague and Influenza? Every war that's every been fought? Hiroshima, Nagasaki, Dresden?

The disturbing implication is that everything bad that happens to us comes directly from God, which makes God into a really horrible bully and tormentor. Bad things happen because shit happens. It doesn't happen evenly—each person will have bad things unevenly spread throughout his/her life, because things don't spread themselves out evenly. Bad things will come in clumps, as will good things.

And some people will suffer much more than others for the same reason - bad things don't spread themselves evenly between people. Of course, poor people will suffer more than rich, and they'll suffer more from the same things that happen to rich people because the rich always have money to ameliorate bad things.

God is not the cause of bad things, unless you deserve it, and even then, justice is uneven and unjust. If you believe in God and have a faith structure that goes along with that belief, then it is likely that you believe that man has a broken relationship with God that has brought evil into the world. It's a free will thing.

Life will sometimes give you more than you can handle. God is the comforter, the companion, the one who walks beside you when you most need it, the one who welcomes you into whatever follows this life.

And sometimes, there's a miracle. But sometimes, you die.

Shit happens. God is good. You gotta live in that tension.

DO YOU BELIEVE IN THE SAYING "EVERYTHING HAPPENS FOR A REASON" AND IF SO THEN WHAT HAPPENS TO OUR FREE WILL AND THE RIGHT TO CHOOSE?

No, unless you want to include the indifferent laws of nature as being that reason.

But if you want to think that everything that happens, happens according to some divine plan and has been choreographed just for you, that's ridiculous.

The reason we want to believe this is that we deeply, passionately want to believe that there is some purpose to our lives and to the terrible things that sometimes happen to us. Fair enough. It's a human need.

But unless there is a God behind it all, then nothing happens for a reason. Things just … happen. Shit happens. You can learn from some of those things, but some of them will kill you. Learning stops when you're dead.

And some of them are so terrible that you learning something is not the point. The Holocaust. The Black Plague. Hiroshima and Nagasaki. War. Earthquakes. Famine. Hurricanes. Tornados. Floods. Drought. A child dies. A spouse.

Then you either survive, or you don't. Maybe humanity picks up a lesson or two, but not as often as we might hope. Witness climate change deniers, anti-vaxxers, racists, misogynists, misanthropes, nationalists, jingoists, and religious fanatics.

God does not have a wonderful plan for your life. You have free will. You get to make choices. They won't all work out. Then you'll have more choices, and they won't all work out.

God does have a plan. It's just hard to figure out. I mean, he's God and we're not. But it seems to have something to do with making a constant, on-going, unremitting choice to love God, and love each other.

Thankfully, there's grace and mercy for us all. And there's meaning and and purpose to be found in the relationship we can have with God who is always with us when shit happens.

Maybe, then, everything does happen for a reason. That reason is to cause you to look to God in the midst of the shit.

HOW DOES THE "LAW OF SUPERPOSITION" HELP TO RECONSTRUCT BIBLE CREATION STORY AND PROPERLY PUT THE FISH, DINOSAUR AND HUMANS IN A PROPER TIME SEQUENCE?

The Biblical creation story is not a history of creation, but a narrative of God's relationship with humans. So there's no science to be drawn from it, just faith.

And the narrative is, God is God and humans are not. Humans tend to elevate other things to the level of God-hood, but those things are not God.

There is one God and only one God, and he made everything. Humanity is not God, and nothing humanity elevates is God. It's not a story about how God made everything. It's a story about that God made everything.

We humans like to worship useless shit. Golden calves, Ferraris, smartphones, rock gods (That is, rock music. Although we have worshipped gods made out of rocks, too). Even our own minds and intelligence.

None of those is worthy of worship. Genesis, in part, is that story. There are other subtleties, too, like prophecies about the coming of a savior. But it's not a story about fish, dinosaurs, and humans being in a proper time sequence.

BTW, the word you might be looking for is "idolatry". That's what we tend to do. God is not a fan.

OTHER THAN SINGING WHAT ELSE COULD YOU DO TO GIVE TO GOD?

Not to be trite at all, but:

If I speak in the tongues of men or of angels, but do not have love, I am only a resounding gong or a clanging cymbal. If I have the gift of prophecy and can fathom all mysteries and all knowledge, and if I have a faith that can move mountains, but do not have love, I am nothing. If I give all I possess to the poor and give over my body to hardship that I may boast, but do not have love, I gain nothing.

Love is patient, love is kind. It does not envy, it does not boast, it is not proud. It does not dishonor others, it is not self-seeking, it is not easily angered, it keeps no record of wrongs. Love does not delight in evil but rejoices with the truth. It always protects, always trusts, always hopes, always perseveres.

Love never fails. But where there are prophecies, they will cease; where there are tongues, they will be stilled; where there is knowledge, it will pass away. For we know in part and we prophesy in part, but when completeness comes, what is in part disappears. When I was a child, I talked like a child, I thought like a child, I reasoned like a child. When I became a man, I put the ways of childhood behind me. For now we see only a reflection as in a mirror; then we shall see face to face. Now I know in part; then I shall know fully, even as I am fully known.

And now these three remain: faith, hope and love. But the greatest of these is love.

WHAT IS YOUR REACTION TO THE STATEMENT "YOU ARE RESPONSIBLE FOR EVERYTHING THAT HAPPENS TO YOU"?

You are responsible for some of the things that happen to you. But if a building falls on your head, that's not on you. Well, I mean, it's *on* you, but it's not your fault. Or your plane crashes. Ship sinks. Government decides to kill you and all your kind. ASTEROID!!! Poisonous snake in your toilet bites you on the willy. And so on.

That's the metaphor. Life is Chaotic, with a big C from Chaos Theory. Tiny little things can make a huge difference, and by definition, they are unpredictable and you can't see them coming.

Getting your girlfriend preggers, eating yourself into sumo-ness, dropping out of school, getting on drugs, robbing a bank, and going to jail. Those are on you.

You are responsible for how you react to things, of course. Ranging from "Is that my kid?" to "Holyyyyy shiiii...!!!" as the building lands on your head. You can be a jerk or a decent human being. Unless a building lands on your head. Then what comes next depends upon whether or not you and God are good. Were. Were good.

HAVE YOU EXPERIENCED A SIGN FROM GOD TO KEEP YOU GOING IN DIFFICULT CIRCUMSTANCES?

Sure, but one man's sign is another man's delusion, so it's a subjective judgment call. Some people see signs from God almost constantly; others would say that those people are finding signs where no signs exist.

So looking for signs could be a self-reinforcing delusion, and it might actually detract from the real experience of having the presence of God with all the time, so that you don't need signs. You already know that God is there. As he always is.

WHAT IS ONE MIRACULOUS THING THAT YOU BELIEVE A HIGHER POWER (GOD) ACHIEVED FOR YOU?

At age 49 in 2002, I started a new venture, traveling the world to give talks to high school students on science, knowledge, and the existence of God. It is a miracle that any school at all ever invites me to come. I cost money, I require time and space, and my talks are challenging to everyone. I've been in over 280 schools in 43 countries now. It must be admitted that I'm very good, but still, it is a miracle.

DO YOU AGREE WITH THE STATEMENT "YOU ARE ONLY AS SMART AS YOU ARE EDUCATED"?

If you expand the meaning of the word "educated" to include all the things that you have learned in life both in and out of school, then you are exactly as smart as you are educated. Well, not quite. You are as informed as you are educated, except only as much as you actually actively remember, use and apply, which is clearly just a small percentage of what you have learned. Whether or not you handle what you remember and apply with intelligence, wisdom and grace is up to you and defines how smart you are.

IS THE BIG BANG IMPOSSIBLE WITHOUT GOD?

It's impossible to say.

We have no science to take us to the moment where the singularity flared into the universe, since all of the science we need came into existence shortly thereafter.

Well, not totally true. Quantum physics could have been there, although there wasn't a there there, and BB could have been a quantum event.

In which case, you've got three possibilities:

It was a random, quantum fluctuation. That doesn't seem likely because

You've got to have an observer in order for anything quantum to happen. So there would have had to have been an observer. God is an obvious choice for that, the leading contender, since the other choice is that the universe existed in a state of superposition for 13.8B years minus a bit until humans arrived to observe it and collapse the quantum state all the way back to BB. John Wheeler and Stephen Hawking liked the latter option. I find it awkward and forced. But they were a lot smarter than I, so maybe.

QM didn't exist yet either, so you didn't have any science at all to make it happen, in which case God is an entirely reasonable solution to the problem. There are no other reasonable solutions, at least none that offer any hope of ever providing evidence.

We should define this God, I suppose. He has two attributes—observer (intelligent, sentient and self-aware) and orderer (since the universe arrived in the lowest entropic state it would ever be in, along with being fine-tuned in all the requisite ways). To that I would add that he is an interactor, since the universe exists solely as a product of interactions, another word for which is relata, relationships between things, and the initial quantum observation was just an interaction between observer and observed; in that case, Singularity and God.

For those who insist that God must be caused or created, those are restricted to this universe, and since God will have created everything about this universe, causation and creation are part of it, not part of wherever God might happen to be, since he is outside of this universe, unbound by time, space, or the laws of physics.

So the most logical and reasonable answer are that God is the most likely source of the universe via BB, absent other possibilities that we haven't come up with yet. None of the other options (multiple BBs, the Multiverse, whatever) preclude the existence of god, though their proponents would like you to think so.

WHAT RELIGION PUSHED YOU AWAY FROM GOD?

Religion has pushed me away from religion. It would be foolish to let it push me away from God. Religion is the tiny box into which man puts his anemic understanding of God. God doesn't fit in that box. "Love God, love your neighbor as yourself" doesn't need a box. Doesn't work as well in a church, synagogue, or temple. Works best out in the world.

HOW DO YOU KEEP FAITH IN YOUR IDEAS?

I have faith in God. I don't consider him to be an idea, though of course those who don't believe would.

As far as having faith in ideas goes, you have to do that pretty gingerly, because ideas turn out to be wrong, flawed, a bit short of truth, and/or dangerous.

True often in science, true in history, true in religion, true in philosophy, true in law, pretty much true in all directions.

So maybe having faith in ideas is much more an exercise in humility than it is a fight for truth.

AS CHRISTENDOM SLOWLY DIES, WILL THE NEXT VERSION OF CHRISTIANITY HAVE A BIGGER IMPACT THAN THE REFORMATION HAS?

If it's just another religion, then maybe, maybe not. It's a crapshoot with religion.

But if God exists and Jesus is who he said he is, then this admittedly sad version of the faith will wither, and the faith will thrive as it always has, but in a different form.

I STARTED READING THE BIBLE AND I'M CURIOUS WHAT PART WILL TALKED ABOUT THE DINOSAURS?

It's not a science book. It's not about science. It's about the history of man's relationship with God. It's not about how the universe was created or how life evolved from simple to complex forms. It's about God saying to man, hey, I made all of this, not any of those other shitty little gods you think are so awesome, and btw, you are doing terrible things to try to keep those other shitty little gods happy, so wake up, pay attention, and let's talk about love, grace, mercy, and forgiveness.

Dinosaurs are awesome. They're just not in the Bible.

IS IT POSSIBLE THERE ARE 2 EARTHS, BUT IN SEPARATE UNIVERSES? THAT WAY GOD HAS A BACKUP PLAN IF WE BLOW OURSELVES UP.

Heeheehee. Sure, why not? What a great question. I mean, I don't think so, but unlike most people who answer questions about God and via their answers reduce him to something about which they can ask "Why would he do that?" in a kind of superior, condescending way. I'm thinking that being condescending about God and assuming that his reasons are trite and idiotic is, well, unwise and ill-informed. Seriously. If the God or gods that you do or do not believe in can be ridiculed, then you clearly need to figure out that God is way more complex and interesting than we could ever imagine. Since we can't even figure out our quantum, relativistic universe, can't figure out dark energy or matter, can't figure out the observer problem, can't really even figure out how gravity works, then we need to shut the heck up about having figured God out.

Having said that, I still don't think so. But Plan B is always a good way to go.

WHAT ARE SOME PAST LESSONS OF OUR HISTORY THAT COULD BE USEFUL TO OUR PRESENT AND THAT WE HAVE NOT APPLIED YET?

A quote I read recently seems apropos (have forgotten the source): What we learn from history is that we don't learn from history.

History, it seems to me, is largely the story of the rich and powerful oppressing the rest of us, and convincing us that it is the poorest and least powerful among us who are the problem, so that we will place our fear, anger and blame upon them rather than on the oligarchs. From time to time, the oligarchs found someone other than that to blame, but it was all to justify doing with us, and convincing us to follow them in doing, what they really wanted to do.

That's painting with a pretty broad brush, but I think you can make it fit.

So in the US (where I live), our history is one of oppressing various and sundry classes of people who were defenseless so that we could hide what was really going on.

The Revolutionary War was the rich, white landowners convincing everyone else to fight for "freedom" when it was really just a revolt against a modest taxation by a distant and distracted King, who quickly became Satan or the AntiChrist or whatever forms of evil we could conjure, and we became the Shining City on the Hill, chosen by God to be his special people. The rich, white landowners, that is. Not the tenant farmers, black slaves, women, or native Americans.

Then the rich white landowners wrote the Declaration of Independence and the Constitution to benefit they themselves, and not tenant farmers, black slaves, or women, not to mention the native Americans.

So we've never actually fought a war for actual "freedom," though all of our wars have been marketed like that. That is, there have been no wars where we fought for the freedom of all Americans, unless you count the Civil War, but half of us fought to keep our slaves while the other half fought to preserve the Union, so it's hard to count that one.

All of our wars have been fought for the benefit of the national interest, usually financial in some way. We'd like to pretend that we fought WWII to defend ourselves against evil, and Hitler was evil, but we didn't fight to protect or save the Jews—we didn't care about the Jews—but really to keep Europe from being taken over by Germany. It was just a colonial war, two sets of colonial powers resisting each other. True with Japan in Asia, as well.

"Freedom" was, and is, the marketing tool used to convince us to go fight and die for oil, in recent years, and demonizing the enemy goes hand in hand with it, along with deifying our cause—we always fight with God against the demons of hell. Such has it always been, regardless of which side is which. All sides claim God's imprimatur on their nationalist escapades.

The argument then might be made that the Germans and the Japanese would have been harsh conquerers, and indeed they were and would have been.

But so were the rest of the colonial powers, including the US, as our history with native Americans will demonstrate, along with our history of supporting despots and tyrants in Asia, Latin America, and Africa as long as they weren't communist despots and tyrants.

So in our modern times, we can see the same thing going on. Many of us love to see the US as bully, forcing the world to kowtow to our will. We are happy to blame refugees, immigrants, blacks, hispanics, poor and homeless people, and liberal Democrats who might as well be commies and socialists.Trump as prez cozied up to tyrants and dictators and pushed our allies around like recalcitrant children. It's all very similar to what happened in Germany in the 1930s—racism, anti-semitism, homophobia, make Germany great again, victimizing the disabled, the socialists and communists, finding a way to justify the gross mistreatment of children and families, and maybe worst of all, finding people reputed to have deep faith and a strong commitment to family values, morals and ethics completely abrogate those beliefs, manipulated by an immoral, narcisstic, egotistical, racist, xenophobic failured bully into blaming immigrants and refugees, the poor and homeless, the people who are different for their fears, angers and resentments, as the oligarchs grow richer and richer and continue to escape any sort of punishment or blame for impoverishing the rest of us.

If that sounds like America right now, or England, Poland, Brazil, or any number of other countries, then once again, many are failing to learn the lessons from history. Those lessons are so painfully obvious, so easily realized, so easily learned. But fear tends to win, and the oligarchs know well how to manipulate our fears.

CAN THERE BE TRUE JUSTICE IN THE UNIVERSE IF THERE IS NO GOD?

If you look around, you'll see little evidence of persistent justice on the earth (what goes on in the rest of the universe is a bit immaterial until aliens show up). So if God exists, then the answer is all around—justice is rarely given fairly, more often doled out on the basis of wealth, social status, and skin color.

It is only if God exists that we dare to hope for justice, and that only in whatever form of after-life might be there, if any at all.

But God is, yes, a God of justice, but he is far more a God of love, mercy, grace and forgiveness.

So if your own faith is one that measures worth on the basis of following religious mores and codes that are random, arbitrary, foolish, silly and often as unjust as the justice you seek, then you will not find that faith, that god or gods whenever you check out of this life and into the next.

If your god or gods compels you to eat this and not that, to wear this and not that, to say this and not that, to do this and not that, then there is no justice waiting for you, because you have failed at balancing out the faith ledger, as have we all.

But if God is a God of love, mercy, grace and forgiveness, then the search is not for justice, because that search ends up with all condemned. The search is for forgiveness in spite of the horrors, the injustice, the pain inflicted on this earth.

You don't want justice, because it is a 2-edged sword that cleaves in both directions.

You want grace, but that grace of a loving God also cleaves in both directions.

We then run the great risk of being angry when those others who have been unjust are not punished, without noticing that our own injustices have been forgiven.

As Chesterton said, grace is the great offense of the faith. It is that which we desperately want and need, but steadfastly refuse to give, or accept, if it means that all are gifted with the same grace.

IF THERE WAS FOUND A POSITIVE PROOF THAT GOD DID NOT EXIST, SO HELL DID NOT EXIST, WOULD CHRISTIANS STILL BE GOOD?

A lot of non-Christians are good without it, so I suppose they would do the same in about equal numbers, after having way too much fun in the first 24 hours. I kid. There would be mass despair, suicides, screaming and running around, weeping and wailing and gnashing of teeth. Or, they would probably say that what is proof now may not be proof in a week or a month or a year, so why give up on the idea? And life would go on pretty much as before.

Here's the other side of the question—if there was proof positive that God did exist, would atheists change their ways? And the answer is pretty much the same—they'd say something like, well, what's proof now may not be proof later, and I don't really want to believe in God, so life would go on pretty much as before.

IN THE LIGHT OF THE ANTHROPIC PRINCIPLE (THAT THE UNIVERSE WAS FINE-TUNED FOR THE EMERGENCE OF LIFE FROM ITS VERY INCEPTION), WOULDN'T IT MAKE SENSE TO SAY THERE WAS AN INTELLIGENT BEING WHO PREPLANNED HUMAN LIFE?

Not to disagree with everyone, but, well, OK, yes, to disagree with everyone. It's surely an option. It is fine-tuned to produce and support conscious life, like it or not, to an unbelievable extent. The only other option that has been suggested is the Multiverse, where in an infinity of universes, this is one variant that got lucky. The obvious problem with that suggestion is that the multiverse is an option without evidence, and without any hope of ever finding evidence. As many have said, it makes far more sense to consider an intelligent source for the universe for which there is ample evidence than to conjure up a mythical host of other imaginary universes.

But talking about evidence for God to be found in science is supremely threatening to many. Not surprising, unexpected, or unfair. Those who believe in God have hardly distinguished themselves on the issue. In fact, they've generally been appalling.

That's not God's fault, though. There's a difference between fearful ignorance (on both sides) and evidence. What fearful ignorance does is ignore actual evidence when the evidence is, ah, disturbing. For both sides.

WHY DOES GOD ALLOW SUFFERING? HOW IS THE THEODICY QUESTION ANSWERED IN THE WORLD RELIGIONS?

The simplest answer is that God allows for free will, and some of the time, free will choices cause suffering. So a better question would be, why did God give us free will when it would inevitably cause suffering?

I don't know, and nobody does. I suppose you could say that without free will, there is no love or hate, no good or evil in the world. We would all just be random by-products of indifferent physical laws in a cold, uncaring universe, with no apparent choice as to how our lives would play out, condemned to suffer whatever happens to us.

Here's another good question, preceded by an observation. We are eager to condemn God for the suffering in far off lands, or the suffering that we ourselves undergo or witness, and we want him to stop that suffering.

But we do not want God to restrict our own personal free will choices, even though some of them will cause suffering.

Fix those people over there, but don't fix me. Don't take away my freedom to choose.

Actually, I don't really need to ask the question. We don't want God to stop the suffering that we each cause, just the suffering that we each experience, or that exotic people in far off lands suffer on a massive scale.

No doubt you will be tempted to say, what about natural disasters—earthquakes, tsunamis, volcanoes, fires, floods, famines, etc.?

And the answer is ... poor people suffer far greater from natural causes than rich people.

And it is human indifference to poverty that allows and perpetuates it.

But but but, you might say, what about when rich people suffer?

Frankly, it seems only fair that they will suffer every now and then.

In terms of world religions, they each have their explanations. Christianity—suffering is caused by indifference, and Christians need to focus on alleviating suffering as part of their journey of faith. Oh, and sin. Disobedience to God. That's on you. Judaism—disobedience to God, and man's inhumanity to man. Islam—inshallah. It's God's will that you suffer, so get over it and obey God. Not that that will help much. Buddhism—life is suffering, suffering is caused by desire, so eliminate suffering by eliminating desire. You can't actually desire anything, including relationships, if you are going to be a good Buddhist. Hinduism—Karma and Dharma. Dharma is duty, so do what you are supposed to do, and you'll get good karma. You are born into a station in life, so you can't aspire to greater things. You can only hope that you'll build up enough good karma to be reborn into a higher station.

If I had to sum up all of the religions as they are practiced, there is an oft unrealized assumption that God is pretty constantly pretty pissed off at us, and we need to spend our time trying to figure out what it is that he wants us to do, and do that, and figure out what he wants us to avoid doing, and avoid doing that.

So we make up a lot of useless ridiculous crap to try to keep God happy, a lot of it focused strangely on constraining women, and sex.

I think, contrarily, that God is a God of love, and he expects us to act as though we love him, each other, and ourselves. That's complicated enough, but attainable, and our inability to balance this out explains suffering on the planet.

IS GOD THE COLLECTION OF ALL HIS CREATIONS?

If God exists, we do not get to define him. We try, lord knows we try, but any definition or description we come up with is going to be horribly, probably infinitely reductive, like the blind men trying to describe an elephant by touch, each man coming up with his own, partial, deficient image.

If God does not exist, feel free to make up whatever you want. It will have no meaning.

BTW, it is inevitable that we will try to reduce God to something we can understand, some metaphor we find revealing. We are tiny and transient, God is decidedly not. So we'll never hit it.

The only way for us to come up with any description of God is for God to offer us one. He could either do that via revelation, which all deistic faiths claim (and either only one of them is right, or they're all wrong), or by showing up. Christians believe he showed up and revealed the parts of his character that he chose to reveal, and then went away, though sort of not completely. Even those parts are hard to understand, which is a metaphor on its own.

Frankly, that's really your only choice. If that wasn't God, then the whole idea of any one faith being true is a crap shoot.

I THINK I AM A GOD. AM I CRAZY?

You're just like everybody else.

The history of humanity vis a vis God is essentially humans wanting to be gods. If you had to sum up the message of the Bible, for example, the best way would be as follows: God is God, and you are not.

We constantly reduce God in size until he is something we can control and manipulate for our own ends, usually something to do with health, wealth, and power. Even if we don't believe in God, the God we don't believe in is usually pretty nasty and pathetic.

So God (if s/he/they exist) has spent eons trying to get us to stop doing that. If I were God, I'd have given up a long time ago and cleaned the universe of our presence.

But, as noted, I'm not God.

THEISTS. IF GOD CREATED EVERYTHING, WHY WOULD HE MAKE PARSNIPS FLOAT AND CARROTS SINK?

He tried doing it the other way, and it was an unmitigated disaster. Atomic structure broke down, galaxies flew apart, Romeo detested Juliet, it suddenly became Catch-42, feces floated up, and Trump became president. Or maybe I have that backwards.

CAN CHAOS EXIST WITHOUT ORDER?

Depends upon what you mean by "chaos."

If you mean "more entropy," that is, more disorder, then no, more chaos comes from less chaos, and if you track it all the way back to BB, then the universe began in the highest state of order (the lowest entropic state) that it would ever be in, and the history of the universe is going from order to disorder, or to more chaos.

If you mean "Chaos Theory," then that's a different question. Chaos is when tiny events have a disproportionate impact on the outcome of a series of events. That can result in a big mess (little c-chaos) or a higher form of order (Fractals, Complexity Theory).

WHEN ONE HAS PROBLEMS IN THEIR LIFE, DO THEY LOOK TO GOD OR HUMAN BEINGS FOR SOLVING IT?

Often people look to God. And often, God looks back at people to solve it. It's a lot easier and more effective for God to work through people to solve problems for other people, esp since that's what he told us to do—love him, love others.

DOES GOD WANT THE SINNERS MORE OR THE SAINTS?

If you're gonna believe in God, then here's the deal. He came for the sinners. There is nobody else. Even the saints are sinners. Fortunately, God is full of love, grace and mercy, so it all works out.

WHEN WILL RELIGIOUS PEOPLE ACCEPT THE FACT THERE IS NO EVIDENCE TO PROVE GOD IS REAL?

There is evidence. Just no proof. But there's no proof of anything in science. Just differing levels of evidence.

Evidence will be accepted or rejected according to one's personal bias. Scientific evidence? Big Bang, quantum theory and the observer problem, fine tuning, the low entropy early universe, and so on. But many will deny all of that, included a lot of conservative religious folk who don't like Big Bang or an old universe, along with a lot of atheists who don't like religious folks or the God they think they understand (the one offered to them by conservative religious folks.)

HOW DID THE UNIVERSE BEGIN WITHOUT THE GOD?

We don't know how the universe began, with or without God.

We know pretty well what happened well inside the first second of the existence of the universe: space and time arrived, the laws of physics arrived, quickly followed by energy, matter, and eventually everything else.

But we don't know how it began, what started it off, where it came from, what caused it to become a universe, and not only a universe, but a phenomenally well-ordered universe with unbelievably low entropy and an astonishing level of fine-tuning.

And we will never know. We'll have thoughts, ideas, conjectures, hypothesis, maybe a bit of evidence hither and yon.

But we'll never know. God is as good a guess as any. Everything came from nothing in a tiny fraction of a second. As you can see from other answers, a lot of folks don't like that much. Their dislike is not based on science or evidence, just a bias against anything that might allow God to have a role to play. That's because religious people have tended to be idiots about the whole thing. Bullheaded ignorance on all sides. Kind of ironic.

CAN PHYSICS PROVE THERE IS NO FREE WILL?

When Pierre Simon Laplace got ahold of Newton's book *Principia*, he got rid of both God (directly) and free will (indirectly). His reasoning was that if the laws of physics were eternal and immutable (in an eternally existent universe), then we had no need of God, and everything that anyone would ever do or had done would be predictable. Thus, what appeared to us humans as free will decisions were in fact just the laws of nature, working.

All of this was predicated on Newton's physics and an eternal universe.

When the GTR came along and showed us that the universe was not eternal (via Big Bang), then the edifice collapsed, because the laws of physics were then not eternal and had not always been working. When Quantum Theory came along and showed us that the universe was probabilistic rather than deterministic, then suddenly the possibility for both free will and the existence of God were back into the conversation.

In addition, the problem of free will (now that physics allows for it) is a brain problem, and projects like Paul Allen's Brain Atlas have shown us that we have little hope of understanding the human brain, in which free will would reside.

So if physics claimed to have been able to prove that free will did not exist, and if that proof has evaporated, then presumably physics can offer evidence that free will does exist, as it has done. There is no proof and never will be, but the latest work in neuroscience seems to indicate that free will does exist. As a corollary, God may also exist, since the proof of his/her/their non-existence has also evaporated.

Isn't physics great?

IN QUANTUM PHYSICS, WHO OR WHAT WAS THE FIRST OBSERVER?

As it happens, most often science moves ahead more quickly than people notice. So for this who persist in thinking that the observer is not, well, conscious, they have not been paying attention:

From Scientific American Mar 2018: "For almost a century, physicists have wondered whether the most counterintuitive predictions of quantum mechanics (QM) could actually be true.

"Only in recent years has the technology necessary for answering this question become accessible, enabling a string of experimental results—including startling ones reported in 2007 and 2010 and culminating now with a remarkable test in May—that show that key predictions of QM are indeed correct.

"Taken together, these experiments indicate that the everyday world we perceive does not exist until observed, which in turn suggests a primary role for mind in nature.

"Now that the most philosophically controversial predictions of QM have—finally—been experimentally confirmed without remaining loopholes, there are no excuses left for those who want to avoid confronting the implications of QM.

"Lest we continue to live according to a view of reality now known to be false, we must shift the cultural dialogue towards coming to grips with what nature is repeatedly telling us about herself."

From New Scientist in 2007: "… experiments led by a group at the University of Vienna, Austria, provide the most compelling evidence yet that there is no objective reality beyond what we observe. Rather than passively observing it, we in fact create reality."

From Margaret Wertheim, aeon. com, Dec 2015: (In) April (2015), "Nature Physics reported on a set of experiments showing a similar effect using helium atoms.

"Andrew Truscott, the Australian scientist who spearheaded the helium work, noted in Physics Today that '99.999 per cent of physicists would say that the measurement… brings the observable into reality.'

"In other words, human subjectivity is drawing forth the world."

(NewScientist Dec 2018): "This raises all sorts of hairy questions. For a start, what counts as a measurement that takes us from probability to certainty?

"Quantum experiments have shown that it seems to involve not just doing something with a measuring instrument, but also consciously noticing the result."

From Steven Weinberg in Scientific American July 2018:

"'Fundamentally, I have an ideal of what a physical theory should be,' says Nobel laureate physicist Steven Weinberg. 'It should be something that doesn't refer in any specific way to human beings…

"'It shouldn't have human beings at the beginning in the laws of nature.

"'And yet I don't see any way of formulating quantum mechanics without an interpretive postulate that refers to what happens when people choose to measure one thing or another thing.'

"But for now, at least, quantum mechanics largely seems to withstand every test.

"'No, we're not facing any crisis. That's the problem!' Weinberg says. 'In the past, we made progress when existing theories ran into difficulties. There's nothing like that with quantum mechanics. It's not in conflict with observation at all.

"'It's a problem of failing to satisfy the reactionary philosophical preconceptions of people like me.'"

So let it be established that a conscious observer is fundamental.

Now, to your question. There are two answers that have been suggested.

One is God as the original observer.

Two is that humans collapsed the wave function backwards in time to Big Bang after we arrived on the scene.

As to who the first human would have been conscious enough to make the observation, that is currently beyond the scope of science to answer, and will likely remain so, since we can't really even define consciousness yet, and will likely never be able to do so. We'll have to leave it as the Quantum Adam, just like evolution has its Mitochondrial Eve.

One and Two are not mutually exclusive.

WHAT STEPS CAN BE TAKEN TO STOP WARS IN THE UNIVERSE?

Keep humans from going there. That should do it. Considering how big space is, and how unlikely it is that we'll ever get out of the solar system, much less to other star systems and planets, I think the universe is safe from wars, especially since the chances of there being aliens anywhere are slim to none.

So we'll stay here and fight our petty little wars over pride and territory, and the universe will never know what fun it is missing.

WHAT SHOULD YOU DO WHEN YOU FEEL LIKE GOD IS NOT HEARING YOU?

Lots of answers here (Quora), some good, some weird. But of course, weirdness is in the eye of the beholder. In terms of you as an individual, I think Mother Theresa said much the same thing throughout her entire life, but she kept on keeping the faith. I would do that. However, there's another story.

The OT is full of examples of God refusing to hear the prayers of his people. always when they were doing evil things but acting as though they were good. There was usually willful disobedience, injustice, oppression, violence, listening to false prophets and so on.

In our more modern times, I find parallels in American Christians who owned slaves, hated blacks, created Jim Crow laws to oppress the blacks, and then went to church on Sunday, sang hymns, prayed, and worshipped.

Similarly, I found parallels these days in Christians who voted for and almost deify President Trump, who want to deport good, decent, hard-working immigrants, who deny the racism in America, who worship wealth and consumerism (and for that matter, capitalism), who sacrifice the freedom and prosperity of others for their own freedom and prosperity, who believe that America is God's chosen nation and deify American things, but really only for white, Christian Americans.

You might be caught up in an era where God is not listening to the prayers of his alleged people in this country because they are caught up in idolatry. God may not hear your prayers because you are one of them.

Only you can answer that.

WHY DID GOD INVENT BILLIONS OF LIGHT YEARS WORTH (AND MORE) OF EMPTY SPACE? AS AN ATHEIST, I WOULD LIKE TO HEAR AN EXPLANATION.

Your question, oddly enough, is amazingly anthropocentric, as though humans occupy the space from which everything can be judged.

It's the same line of thought that the church had when it declared that since God created humans and the Earth, they must occupy the spatial center of the universe—the geocentric universe. Which was absurd.

And it's the same line of thought, conversely, that science had when it said that since humans and the Earth aren't the spatial center, humans and Earth don't matter at all. Which was and remains absurd.

The size of space relative to the Earth says nothing at all about the importance, or lack thereof, of the Earth, or of the humans on it. Carl Sagan made the same mistake in Contact—"it seems like an awful waste of space."

Indeed it might—to us. But we are not God. Sagan's character Ellie in Contact said it much better—"we are tiny and insignificant, and rare and precious."

That's us to God. So tiny in time and space as almost really not even to have been here at all, but rare and precious enough for whom to sling a grand tapestry of cosmos into existence.

So how can we say such a thing? The nature of the universe itself is our finest clue—Big Bang, everything coming from nothing via no science for no apparent reason in virtually no time at all, the laws of physics arriving and choreographing our own arrival long long after, the universe finely-tuned to such exquisite precision as to cause scientists to weep and gnash their teeth in frustration, and capped off with the increasingly inescapable need for intelligent observers to weave the strands together into a glorious cosmic web.

It may be immense, and we may be immensely small. But size, in this case, both matters and does not. It's a quantum thing. It is meant to show us two things; that we are tiny and insignificant, but are made rare and precious by the artist himself. Or her. It. Them. It hardly matters. Your mistake, and that of all us, is a massive failure of imagination. In that, size does indeed matter.

WHAT DOES GOD SAY ABOUT SOUL MATES?

Nothing. No such thing. If you and someone else feel like soul mates, great, that's not a God thing. It's not even a thing. It's just deep love, which is super, but don't get wrapped around that axle.

IF GOD CREATED EVIL, AS IS STATED IN ISAIAH 45:7, WHY DO CHRISTIANS INSIST HE IS "GOOD"?

Not a good interpretation of that verse, for starters. There's prosperity theology, which is, frankly, heretical, and there's disaster theology, which is not. God does tell the Israelites that if they obey his commands, they'll prosper as a people, but this is not a general message to everyone, clearly, since there are many believers to whom terrible things happen, and many terrible people to whom wonderful things happen. It's all a bit more complicated than that.

But disaster theology is a whole 'nother thing. Isaiah, Jeremiah, Ezekiel, Amos, they're all full of disaster theology, which works like this—do the right thing, or I'll land on your head with both feet. What's the right thing? Love God and love your neighbor as yourself. So who's my neighbor? Anyone you find in need of help. Black. White. Green. Poor. Homeless. Rich and depressed. Disenfranchised. Refugees. Immigrants. Straight. Gay. Prostitutes. Drug addicts. Drunks. Americans. Not Americans. Capitalists. Communists. Socialists. Anyone. Everyone.

And if you don't do that, then disaster theology is barking at the door.

Now, evil. That's something else altogether. Oh, wait. It isn't, really. Evil happens when people choose not to love God or love people as themselves. See above.

So God gives you free will, and you can make a choice. Be a jerk. Love yourself first, last and always. Or love God and love others.

So now you know what's good, and what's bad. God wants you to do good and not bad. But you, and I, and everyone do enough bad to create evil in the world.

Then, you will say, what about Earthquakes and tsunamis and floods and hurricanes and all other manner of natural disasters that kill lots of folks? How is that my fault?

Well. If you were paying attention, you would notice that many more poor people are killed by natural disasters than rich people. And they are poor because the rest of us make them poor.

OK, fine, you might then say. But what about death for anyone, not just for poor people, but everyone?

According to Christian (and Jewish and Islamic) scripture, death came into the world because Eve chose to define for herself (and all of us) what was good and what was evil, taking that away from God, becoming the arbiter of her own morality and ethics. Which is what we all do. And in so doing, we seek to become like God, replacing God with ourselves, becoming our own idols, the focus of our own idolatry. That's bad. You don't have to like it. But even then, that's what you are doing. Defining good and evil for yourself.

But, you could then say, what kind of a God would create a universe with the potential for such vast evil within it?

And the deeply unsatisfying answer is, if God does not exist, then there is no definition of good or evil that is true for everyone, every place, every time in history. We then get to define good and evil for ourselves, and that's when the trouble starts.

So without God, nothing is good, nothing is evil. Everything just ... is.

You may not like it, but you have to live in tension with a God who allows evil, because otherwise, nothing in fact is evil at all. You can't call anything evil, because frankly, who the hell are you?

Not God. That's for dang sure.

IF YOU MET GOD, WHAT WOULD BE THE FIRST THING YOU'D ASK HIM?

"Wow, you look taller on TV." Not a question, more of a joke. He'd laugh, tho. He's like that.

ARE WE CONFUSING RELIGION WITH CULTURE?

Religion is inevitably entwined with culture, sometimes in benign ways, sometimes in destructive ways, and needs to be distinguished from faith. Faith is what we have in ordinary things like elevators and airplanes, that they will work the way they are supposed to work, even though we know that sometimes they don't and things get messy.

Faith in God is similar—we know some things to be true, but many of the most important things we have to take on faith, things unprovable. If there is a God (and I believe there is), then for some reason he has set it up this way, giving us a choice to accept or reject the evidence we see, hear, or read about.

Religion is how we express that faith, each one of us in different ways, each religion in different ways. And since we are born into, raised within, and live in different cultures and times, even different versions of each faith will express things and practice things differently.

I would have to guess that none of them are completely correct, and few are completely wrong.

It is when a religion as an expression of faith loses its grip on that faith, subsumed by culture, that things can get really messy.

My background is American southern Christian, so I can easily find examples—somehow our faith in a loving, peaceful Jesus focused on the poor and the oppressed lost sight of that with Indians (i.e. native Americans, the indigenous or aboriginal peoples), slavery, racism, immigration, women, the poor and oppressed, war, capitalism, individualism, even freedom and (lately) democracy and American Christians became inordinately fond of white male landowners, that is, the rich and powerful.

But you can do this with any religion in any country. It's the same old story. Culture, patriotism, nationalism, parochialism, misogyny, misanthropy, racism, greed, power, wealth and many other things can infect authentic faith and drag it into a hideous abyss.

But just because religions tend to do this eventually and inevitably does not negate faith and has little to do with God, who, one would imagine, is no longer shocked but is still appalled at the things that we do in his name to each other. Now, why he lets it go on and on is a good question. But as you cannot blame your parents when you stab your little sister with a fork, well, there you go.

IF GOD IS SO POWERFUL, THEN WHY CAN'T HE PROVE IT, USE IT, AND TO SHOW IT TO US HUMANS?

He gives you the choice. And really, if he did show up as Jesus, acting like and doing all the things that the Bible says and that Christians believe, even then most people did not believe. So because you have a choice, you get to deny God even if he shows up, 10,000 feet tall and looking very much like God would look. As soon as he went away, many would deny that he was here, or call it mass delusion, or whatever. It's the name of the game—free will. The trick is, nobody is smart enough or knowledgable enough either to believe or disbelieve. There's always somebody smarter out there who either disbelieves or believes. Everybody looks at the same evidence, and makes their own decision. Makes it all kind of messy, but there you go.

DO YOU AGREE THAT WHEN WE SPEAK SOMETHING TO THE UNIVERSE THERE IS SOMETHING LISTENING IN THE SPIRITUAL REALM?

Only if 1) God exists and 2) he cares about each individual enough to 3) listen.

The scientific evidence that God exists is strong—Big Bang, everything coming from nothing in a tiny fraction of a second, the fine-tuned universe, the measurement/observer problem in QM, spontaneous emergence and self-organization in Complexity Theory, the return of Lamarckian evolution (evolution with intent and specificity in response to environmental challenge) and the arrival of epigenetics, and so on.

The evidence that God cares is pretty strong, especially with the proposal in recent years that the universe is nothing but interactions. That implies that God is a God of interactions, of relationships, and the highest form of relationship (I propose) is love. So the evidence in nature suggests that God is a God of love. God is love.

What that means is complicated, as love always is. It doesn't mean that humans are spared from suffering, for example, or death. The vast majority of suffering is human on human, so it's a bit presumptuous to blame God for that. Death is a problem, but Islam, Judaism and Christianity have an answer for that, and it's ultimately humanity's fault for presuming to define good and evil for ourselves apart from God, and thereby usurping the role of God in morality and ethics; an act of idolatry, and all three faiths condemn idolatry.

So, then, what is the purpose of speaking to the universe? None whatsoever. But is there a purpose in speaking to God? That is, a purpose to prayer?

Only if he exists, cares, and listens. Believers would say that he does.

What he might choose to do after he listens is an interesting question.

He does not save all souls, relieve all suffering, cease all pain, reward all good behavior, punish all bad behavior, fix all problems, heal all diseases, or provide for all needs. Sometimes he does. Sometimes he doesn't.

So maybe prayer is not a to-do list for God, or a wish list as though he is Santa.

Maybe it's a constant, on-going conversation with the God who loves you, a continual interaction, the life-long pursuit of relationship. I can live with that.

IF GOD CREATED EVERYTHING, WHAT CREATED GOD? COULD IT BE OUR MERE MORTAL MINDS ARE TOO UNSOPHISTICATED TO UNDERSTAND THE COMPLEXITIES OF THE ALPHA AND OMEGA?

Creation happens to and inside this universe. God is therefore outside of this universe, outside of the constraints of spacetime and the laws of physics, outside of cause-and-effect, outside of everything we puny universians understand, outside of the need to be created.

He is inaccessible to us, incomprehensible to us. He can only be accessed as he chooses to make himself accessible, understood as far as he chooses to reveal himself to us.

That's why Christ is such an intriguing figure.

You, and we, have made God too small. We try to contain him within this universe, and ask questions about him as though he is like us, only bigger. If that is true, then he's not really God, and something else would have to be.

IF GOD CAN DO ANYTHING, WHY CAN'T HE SIMPLY MAKE ME GOOD?

For some reason, your free will is more important. The choice to love or to hate, to hurt or to heal, to be kind or cruel, selfless or selfish is what makes us close to both angels and demons. Otherwise, we are just puppets, and for better or worse, that's not what God wants. Interesting how we blame God so often for the ugly, and take credit for the beauty.

COULD GOD BE POSITIVE THINKING AND THE DEVIL BE NEGATIVE THINKING?

Only if God does not exist. Then you can make up whatever you want. I'm betting on God, tho.

IS THERE HISTORICAL EVIDENCE THAT CAN PROVE RELIGION IS REAL?

There's ample evidence that religion is real. Evidence is not proof, but in this case, it's as close as you would want to get.

There is no proof of divine beings, but to say there is no evidence is to conflate the word evidence with the word proof. Of course there's evidence. It's the choice of each person to accept or reject the evidence.

I don't, for example, accept the Shroud of Turin as evidence of anything. It's interesting, it's seems to show the face of a man, but to accept it as evidence that Jesus was crucified is a bit of a stretch. Nor do I accept the vast numbers of ancient relics as evidence of anything. I read somewhere once (long before the Internet, so it must be true :)) that there's a museum that has two skulls side by side on shelf, one small, the other large. The large one is tagged as the skull of, I don't know, Paul or Peter or some Biblical luminary. The small one is tagged as the skull of that luminary as a child.

I do, however, accept evidence found in nature as indicating the existence of God—Big Bang, fine-tuning, the laws of physics, quantum theory. Others do not. That's the nature of evidence.

And I do believe in that Jesus fellow. I accept the historical evidence, the written evidence, and my personal experience as real. I think most Christians have their contemporary heads shoved up somewhere wet, warm, and smelly, but they're humans just like everyone else, subject to being an idiot on regular occasions.

Religion is just man's attempt to try to contain an uncontainable God. We usually do that by coming up with more or less ridiculous rules that we think God wants us to follow. That's because we think God is pissed off, and we're just trying to keep him happy.

That's why I like that Jesus fellow. Love God, love each other, grace, mercy, kindness, forgiveness. Nothing wrong with that.

IS IT REALLY TRUE THAT ALL THINGS ARE POSSIBLE WITH GOD?

All things that God will make possible are possible with God.

On this Earth in this universe, then, not all things are possible. He has put it together with the laws of physics, and unless he chooses to supersede those (which he could do), we are bound by them.

Given the quantum nature of things, strange and miraculous things become possible, but unlikely. The probability of macro objects being in two or more places at once, of quantum teleporting from here to there and so on is very very low, but it is not zero.

On a larger scale, you cannot fly on your own, without the aid of an aircraft of some sort. You cannot fall very far without dying, without the aid of a parachute or something much like it. You cannot run very fast without some sort of aid, nor jump very far. There is an upper limit to your intelligence. And so on.

It's part of the plan that we are limited in our abilities, because the underlying message is that God is God, and we are not. We'd like to be God, but we can only pretend at playing gods, and it's pretty pathetic.

WHY DO EVANGELICALS HAVE A DIFFICULT TIME ACCEPTING THE BIG BANG VIEW ON CREATION VERSUS GENESIS' VIEW?

As a person of faith and a science lecturer in high schools globally for 30 years, let me give it a shot. BTW, I'm a big fan of the Big Bang (as well as the TV show!) and the Big Guy upstairs, and a big fan of Genesis.

American Protestants, because of evolution and its apparent challenge to the creation story, have grown distrustful of science in general, as you might have noticed. They are anti-Big Bang, anti-Climate Change, anti-vaccinations, anti-evolution and so on.

They were told by Bishop Ussher and Martin Luther that Genesis implied (note: implied, not said outright) a divine moment of spontaneous creation about 6000 years ago when, in a 7–24-hour-day process, just like Genesis seems to describe, God spun up the whole of creation out of whole cloth in a scientific whoosh. I was raised to believe this.

Evolution as a theory was seen then as a great challenge, since it took place over immense periods of time. In a 6000-year-old creation context, this was not viable. That we, made in the image of God, would be considered as being descended from the same source as modern apes was an appalling concept.

Darwin's relationship with God was complicated at best, but he did say that it was possible believe in both evolution and God. He didn't seem to figure out how to reconcile Genesis and evolution.

It did not help that the Catholic Church had hung its hat on a geocentric universe, that the Earth was the center of the universe, and had been pretty insistent about it with people like Galileo, Copernicus, and Bruno.

It also did not help that when Newton's physics was released to the world in Principia, thinkers and philosophers decided that Newton's universe needed no God nor allowed for free will.

Science had overreached itself, claiming intellectual territory that was not its to claim, that is, that since we had discovered the laws of physics and biology, we no longer had a need for God to explain anything. They used the God-of-the-gaps argument, that religious folk had spend centuries explaining the once inexplicable by saying that since whatever it was was mysterious, that God must have done it miraculously.

Neither science nor religion seemed open to the idea that God might have done it all via the laws of nature. And thus the chasm was opened between the two. Science described the origins of the universe, the Earth, and life, which was great, but then went a bridge too far and claimed that God did not exist and that religion was ridiculous. Religion refused to bend on the science of origins, and thus ended up looking as ridiculous as science had claimed.

I'm painting both with a broad brush—Darwin, Galileo, Copernicus, Bruno, and Newton were people of faith, though a sometimes complicated faith, as are many well-regarded scientists today, like Francis Collins, John Polkinghorne, Simon Conway Morris, Charles Townes, and others. There are many ordinary believers (like myself) who are believers not only in God but in the promise of science.

Genesis, btw, was never written as a book of science, but as a book introducing God's relationship with mankind. The creation story intends to say, hey, all those puny little idols you guys are worshipping as though they are gods? I am God. I made it all. I created it all.

Also interesting is that when Big Bang appeared on the scene, it was rejected by most scientists, many of whom thought that it sounded way too much like religion and religious creation stories, specifically the one in Genesis. They thought that it was Christians trying to regain the ground they lost because of Newton's physics. Didn't help that the physicist who figured it all out (Big Bang, that is) was also a priest—Georges Lemaitre.

Scientists rejecting BB because it sounded too much like religion, and the religious rejecting it because it sounded too much like science. Irony.

God made the universe, made the laws of physics that made everything happen. The timeline is really not that big of a deal. Evangelicals just need to get over it. God did it. Science just needs to get over it.

WHY ARE RELIGIOUS PEOPLE SO IMMORAL?

Religious people are the only ones who can be immoral, because they have moral absolute standards against which they can be measured.

Irreligious people can move the moral goalposts whenever they like and redefine morality for themselves. So technically they are never immoral, except if they decide that they are.

Both groups act they way they do because they are humans, and humans do bad things sometimes. Well, religious people do bad things as well as good things. Irreligious people just do things—Richard Dawkins tells us that there is no good or evil in a pointless universe—so they get to decide for themselves about good and bad. That, plus laws, courts, jail, that kind of thing. But laws, courts, and jail are notoriously inaccurate in history at measuring good and evil, since many more poor people go to jail than rich people, and black people than white. Which sucks.

Also, religious people who are immoral are so much fun to write about. So they probably get written, talked and reported about more often than irreligious people, since who cares if they get caught? There's no irony in that.

HOW WAS FAITH CREATED?

Faith is a necessary part of human existence, one that we all share. Faith is based on partial knowledge, experience, shared narratives, and ultimately becomes one's autopilot.

For example, much of our daily lives is faith-based. We buy and eat food having faith that it will be good, good for us, and safe, that it won't make us sick. We do this even though every now and then, it isn't good, good for us, or safe, and does make us sick. Still, we buy it and eat it.

We ride elevators the same way, and airplanes, cars, bicycles, trains. We've ridden them before, we know people who've ridden them before, we've seen them work, and so we get on without a second thought, even though we are fully aware that sometimes elevators, planes, cars, bikes and trains crash and kill people.

We breath, having faith that there will be air. We drink, having faith that the water will be safe, unless we are someplace where it isn't safe, and then we take faith-based precautions. We go to the doctor in faith, to the dentist, to have surgeries, take medicines, brush our teeth, use deodorant. The list is endless, because we have to have faith in so many things in order to live our lives without freaking out about it all the time.

We have faith in science, because it has produced so many wondrous things, even though every now and then science gets it wrong—thalidomide, tobacco, opioids, any diet advice ever given.

Some have faith in God in the same way—partial knowledge, experience, shared narratives, historical and cultural and/or sociological traditions, and revelations. But faith of any kind and type is always based on partial knowledge, not full, so there is always faith in something that cannot be known. Many would deny faith in God because they demand to know everything before they might believe, even though they have faith in elevators while knowing very little about the next elevator they get on, or faith in Cheerios without knowing anything at all about how this box of Cheerios came to be made.

The difference, of course, is that God is not seen, and Cheerios are, and Cheerios don't make profound demands on one's life, unless they are somehow toxic or poisonous.

The existence of God is maybe the most important concept ever, because it's radically not trivial or ignorable. So it is often more convenient to act as though God does not exist, because it's easier that way. Or it seems so.

At this point, you'll probably say that there's no evidence for God. In fact, there's lots—Big Bang, fine-tuning, order and structure in the universe, the profoundly interactive nature of the universe, the need for evil actually to be evil, the need we all feel to be loved and to matter—but there's no proof, as there is no proof of anything, only evidence, and evidence is always ignorable and often interpreted according to one's bias. To say that there's no evidence is a faith-position. To say that the evidence is not compelling enough is fair, but inevitably biased. To say that people who believe in God all seem to believe in a different god is accurate and true, but negates neither faith or God's existence. It just makes it all that much more interesting and complicated.

WHAT'S THE BEST THEORY FOR AFTERLIFE IN ALL AROUND THE WORLD FOR YOU?

For me? That's your question?

Science has no theory of an afterlife, nor anyway to collect evidence to support such a theory. And science is all about evidence.

Religions all have afterlife stories, but all the evidence is based on faith.

I believe, I have a personal faith, and it includes an afterlife.

Having said that, I no longer have any idea of what that afterlife might be like. When I was younger, I was taught various things about the afterlife, but there's no way for me to know whether or not they are true, and the Bible is a bit vague and poetic on the topic. So I believe that there is something, but I don't know what it's going to be like.

What most people (e.g. those who don't have a personal faith and belief system based on one of the world's religions) believe is that there is nothing after life. They also tend to believe that we can believe anything we want, because nobody can tell them what to believe. They get to make a choice, and they choose not to believe.

I don't, then, understand their hostility towards people of faith. If God does not exist, then religion is ridiculous, and I get that, but if God does not exist, then anything you might choose to believe in is ridiculous. Everything, to use Richard Dawkins' language, is a delusion; not just religion, but everything.

So humanism is just as much of a delusion. Humans don't matter to the universe, so just because they matter to us, who are humans, has no meaning and is a bit anthropocentric besides.

So if nothing is true and everything is a delusion, then you might as well believe in an afterlife. If it turns out to be true, you'll know and that'll be awesome. If it turns out to be false, you'll never know anyway, no matter if you chose to believe in it or not. No one will ever know.

So the only thing that holds any potential for awesomeness is to believe in God and an afterlife. Like Pascal's wager, you've got nothing to lose, and you might as well, because either nothing is true and everything is a delusion, or the only thing that is not a delusion is that God exists. If he exists. The only thing that you can choose to believe in that might be true is to believe that God exists.

WHY DO WE BELIEVE THAT THERE IS ANY CREATOR OF THIS UNIVERSE?

As you can see from the answers to your question already, everybody has their own opinion and evidence to draw upon. I believe in God as creator, and I believe that Big Bang, evolution, and so on are the methods he used.

But many do not, and the difference of opinions will be wide and full of passion. I believe there is profound evidence of God as creator to be found, that science has found—Big Bang cosmology, Quantum Theory, the CMB, fine-tuning, and more—but many others will look at the same information, the same evidence, the same science and say that it is not evidence of God's existence.

So it's really not about evidence or information; though those cannot be ignored, they cannot be seen as definitive. It's about faith in things outside the evidence that cannot be proven, and each will have his or her own faith structure, his or her own belief system.

Apparently that's the way God intended for it to be. I'm a follower of Isa, you may not be, others certainly are not.

So we need to give each other grace and mercy and space in which to believe as we want. Otherwise, we fight and kill each other and validate the beliefs of all those unbelievers who say that religion is evil, bad for humans, bad for the planet.

We each need to spend much less time defending our faith (frankly, the God who created the universe doesn't need the help) and more time practicing our faith, especially the parts of loving others, caring for others, serving others, and, I would suggest, caring for the planet on which all of the others live so that they might all continue to live there.

We need to spend less time trying to force others who believe differently than we to change their beliefs and become like us, and more time representing our faith is a positive, affirmative way.

It would be so nice if all of those who hate religion could look at religious people and say, well, religion is stupid, but those religious people are so nice. Instead of love, we are far better at demonstrating hatred, fear, anger, and violence.

We should stop that.

ACCORDING TO THE BIBLE, MANKIND HAS BEEN ON EARTH FOR ABOUT 5,000 YEARS OR SO, WHICH SOUNDS ABOUT RIGHT TO ME. HOW AM I WRONG?

The Bible says nothing about mankind having been on the Earth for about 5000 years. That's from Bishop Ussher and Martin Luther, I believe, and is an extrapolation based on incomplete evidence and poor assumptions. So the first place you are wrong is to assume that is from the Bible.

The second place you are wrong is in saying that it "sounds about right" to you. That has no meaning in science or theology. The universe as you look at it is nothing like the way that it really is. And God as you think about him is nothing like the way he really is.

The third place you are wrong is in assuming that the Bible was ever intended to be a book of science. It is a book about the interaction between God and mankind. The creation story is the first part of that interaction, and its overarching intent is for God to tell us that he alone is responsible for creation, not a pantheon of other gods or myths made up by humans. God is God, and you are not. That's the message as it starts, and it remains the message throughout.

CAN YOU BELIEVE IN GOD AND SCIENTIFIC THEORIES LIKE THE BIG BANG AND EVOLUTION? DO SCIENTISTS ABANDON RELIGION OR DOES IT ALLOW A BELIEF IN GOD TOO (AND WHAT FAITH DO THEY TEND TO BE)?

I believe in God, Big Bang, and some version of evolution (though my version has been updated by recent work in epigenetics, Larmarkian evolution and Complexity Theory). Some scientists abandon faith; some acquire faith. And I don't know what version of the faith they tend to choose.

There are those who will say that no question in science can be answered by referring to God, and that science and faith should operate in different magesteria (to use Stephen Jay Gould's word). It is true that the laws of physics either answer or have the potential to answer many if not all of the question we throw at them, but since Big Bang arrived on the scene, the whole picture has changed, so that we now have a very interesting question —where did the laws of physics come from? They arrived after BB in the universe, so that BB was not caused by the laws of physics, apart possibly from quantum mechanics. But that doesn't help, because it is increasingly shown to be true that nothing happens without an observer, so even if BB was a quantum thing, God had to be there to observe the universe into being.

And the laws of physics have revealed to us the finely-tuned universe; whence fine-tuning? More and more it is reasonable to offer God as an answer as to why the universe is the way that it is; not mandated, but reasonable.

So there are those scientists who are moved to believe in the face of evidence uncovered by science itself. And others who refuse.

It's not really about the evidence for the latter; it's about obstinacy, bias, presupposition. They are hard-core Copernicans, but even the Copernican Principle has turned out to be an unwarranted assumption.

Belief in God then turns out to be an evidence-based choice that some choose to refuse to make. That'll generate some heat.

WHAT ARGUMENT MADE YOU STOP BELIEVING IN GOD?

Evil and suffering on Earth. But then I realized (thanks to Richard Dawkins) that if

God doesn't exist, then evil isn't evil (and good isn't good) and suffering isn't suffering, it's just more or less levels of arbitrary and random comfort.

So if God doesn't exist, then I have no way to be mad at him for evil and suffering. It's just shit happening randomly in the universe that is sometimes pleasant, and sometimes not.

It's only if he exists that I can hate evil and suffering.

And it then makes no sense to be mad at God, since he is, after all, God, and being mad at God is a bad business plan.

So I figured out 1) that most evil and suffering is caused by human on human indifference and cruelty, and 2) that's a free will thing. Plus, 3) God didn't promise to keep us from suffering, he just promised 4) to be there with us no matter how dark and desperate things get even if 5) we're all mad at him and stuff.

So I never really stopped believing. I just had to think more clearly about it.

WHAT'S THE ONE AND ONLY REASON THAT WOULD FINALLY LET YOU ACCEPT THE EXISTENCE OF GOD?

There is nothing that will convince all people. If God showed up, hung around awhile, and then left, some would believe, some would deny, and as time passed and the event receded into history, some would deny the history, the stories, reducing them to mass hysteria, psychoses, drugs and alcohol, mythology, mistaken identity, lies.

That is, of course, the case with Jesus Christ. Some believed, some denied, and as time has passed, some have denied that he was ever here in the first place. Some accept that he was here, but as man only, not as the son of God, or as a charlatan, a fraud, a rabble-rouser, a terrorist or revolutionary.

What is true? Either we'll all eventually know, or none of us will ever know. Such is the nature of death. It is either final, or it is a doorway.

DOES GOD EXIST?

If you believe in God, and if God exists, then yes.

If you don't believe in God, and God exists, then yes. If God doesn't exist, then it doesn't matter what you believe. Obviously, if God exists, then it doesn't matter what you believe.

Let's talk about the options.

If you don't believe that there can be any evidence for the existence of God to be found in nature, it still doesn't matter what you believe. If God exists, you were wrong (unless God didn't put any clues to his existence in nature, in which case, you were right in the battle, wrong in the war). If God does not exist, you were right, (unless there are in fact things in nature that could be taken as evidence for the existence of God, even if God doesn't exist, in which case, you were wrong in the battle, right in the war.) Let's take the most interesting case. If God exists, and if there is evidence for that to be found in nature, then to have said that there can be no evidence for God, or that there is no evidence for God yet, is just bull-headed obstinance.

There are many who will say those things. Not only is there no evidence for God, there cannot be evidence for God in nature. Anything we find that might be interpreted as evidence, cannot be evidence.

That's clearly a Flat Earth argument. No matter how much evidence there is that the Earth is not flat, they refuse to believe that any of the evidence is actually evidence, that it is untainted by bias and conspiracy.

If there is evidence for God to be found in nature, then it must be considered and measured in a rational, reasonable, scientific sort of way. It should not be rejected out of hand because of an ignorant, Flat Earth-style argument or a dated God-of-the-gaps argument.

So here's the evidence: Big Bang, just for starters. Then the universe that was produced by BB is both fine-tuned and observer-dependent. Note: This is the evidence that the evidence deniers will deny is evidence. They confuse evidence with proof. This is evidence to be considered, accepted or rejected. There is no proof of anything in science—it's all about evidence. It's either crappy evidence, OK evidence, compelling evidence, or overwhelming evidence, and all the gradations in between.

So the question is, what kind of evidence is this?

The best way to evaluate evidence is not to look to the people who are biased in that direction already, but to people who are biased in the other direction, against God, as it were.

So. How did scientists react to the idea of Big Bang? Just for starters. Here's a couple of representative comments of historical note, widely remembered:

Sir Fred Hoyle, British Astronomer: "The passionate frenzy with which the Big Bang cosmology is clutched to the corporate scientific buxom evidently arises from a deep-rooted attachment to the first chapter of Genesis, religious fundamentalism at its strongest."

Physicist William Bonner: "The underlying motive is, of course, to bring in God as creator. It seems like the opportunity Christian theology has been waiting for ever since science began to depose religion from the minds of rational men in the 17th Century."

George Smoot, Nobel Laureate for COBE: "We have observed the oldest and largest structures ever seen in the early universe. These were the primordial seeds of modern-day structures such as galaxies, clusters of galaxies, and so on. If you're religious, it's like seeing the face of God."

Clearly, many scientists saw BB as evidence for the existence of God, and many rejected BB on that basis, at least until BB itself was overwhelmingly validated by observational evidence. Einstein himself rejected BB initially, though whether or not that was on religious or scientific grounds alone or together is not clear. Worth noting that the guy who came up with the idea of the universe having a starting point at all was a Catholic Priest/Physicist named Georges Lemaitre, whom Einstein initially ridiculed but later praised.

Evidence. Not proof. But pretty good evidence. More evidence could be seen in the rejection of BB by many people of faith because it didn't match their Young Earth leanings.

You get to accept or reject this on your own, of course. It's evidence, not proof. And you may think it's really lousy evidence. It's not no evidence, because others saw it as actual evidence, others who have much more scientific credibility than you are likely to have, no offense. So if you decide it's not evidence at all, then you are just being bull-headed.

Fine-tuning is next. Fine-tuning is the observation that the constants and values associated with the laws of nature are so carefully balanced that it seems very much as though they've been dialed in by, well, a dialer. A creator. They've led to the Weak and Strong Anthropic Principles from physicist Brandon Carter. The Strong version (SAP) is the most interesting:

"We must be prepared to take account of the fact that our location in the Universe is necessarily privileged to the extent of being compatible with our existence as observers." (weak version)

"The Universe (and hence the fundamental parameters on which it depends) must be as to admit the creation of observers within it at some stage." (strong version)

This is tied into the universe being observer-dependent, so we might as well talk about them together. The crux: the universe has been carefully fine-tuned to produce observers, without which we would have nothing in the universe.

So the evidence is strong that there is a grand architect to the universe, someone we might call God, who is outside of time, space and the laws of physics, since BB caused all of those to come into existence. Everybody gets to consider a bigger God than they thought, whether they accept or reject the idea of God. Atheists have been rejecting a far less interesting God than our universe suggests.Are morals individual or universal?

If there is no God, neither. Unless you would accept somehow that an indifferent universe has a universal moral structure without God being involved.

If there is no God, then morals emerge from social interaction at every level.

You might have what you consider to be individual, personal morals, but those emerge from all of your social interactions. No interactions, no need for moral constraints.

Every relationship develops its own moral structure, though very likely not all that different from that of large groups. Neighborhoods do it, villages, cities, counties, states, regions, countries, collections of countries.

Societies are not especially good at evolving morally—witness the Holocaust, slavery, the Killing Fields of Cambodia, the Spanish Inquisition, the Turkish genocide of the Armenians, the genocide of native populations by invading forces in the Americas and Australia (for example), cultural misogyny, child labor, child marriage, Jim Crow laws in the US, anti-immigrant and -refugee actions and laws in the US and elsewhere, the exploitation of workers throughout history, and more. So we can't look to societies either historically or in the present to give us moral guidelines. We certainly can't look at individuals—witness Donald Trump, Harvey Weinstein, and many other appalling examples of personal morality gone sideways.

So either morals are entirely derivative and emergent, or there are absolute moral standards that exist.

Even if we believe in God, we each believe in God in a different way, so we derive our own morals from our own (often) peculiar belief system and then try to force them upon the culture at large. Every religion has it own mostly bizarre set of beliefs that generally are focused mostly on what women can wear and do (better said, not wear and not do), who you can and cannot sleep with, what you are and are not supposed to eat, and so on, but allowing all sorts of forms of mass slaughter, slavery, racism, jingoism, warfare, and wiping out of the infidels.

So here you go—you get my opinion. I believe in God, and I believe in Christ, not least because I like what we were told to do to be moral—love God, love your neighbor as yourself.

I think that's universal, God-given, works all the time, is sufficiently simple that simple people can get it, morally complex enough to confuse philosophers and theologians, is amazingly easy to say and amazingly hard to do, is a constant challenge to everyone, and is something that, if we imposed it by law on every culture, wouldn't piss anyone off and would make the world a better place. It's also delightfully ambiguous and open to interpretation.

Put it together with grace and mercy and you've got a bang-up religion going; kind, forgiving, demanding in a good way, challenging in a good way, individually self-regulating, and holds everyone to a standard they cannot possibly meet, but will always know when and where and with whom they have fallen short.

Don't forget grace and mercy. Forgiveness. Generosity. Compassion. Sacrifice.

Dang. We should just do this thing.

ANSWERS TO QUESTIONS PEOPLE ARE ACTUALLY ASKING

END PARTS

So. Thanks for reading all the way to the end. That greatly exceeds my expectations. If you have violent reactions, disagreements, nausea, or intestinal distress over any of my answers, you may write your own book. You could also look me up on Quora, find those questions and answers, and write passionately about why you think I have my head stuck in a wet, dark, stinky place. Get in line. You'll find lots of others who have already done that. I might even respond. You could read the others and my responses to them and save us both a lot of time.

I strongly encourage you to buy all of my other books. What, you thought this was the first one? Hah, is what I say to that. Here they are:

Science-y books:

Life, the Universe and Everything by Andrew Fletcher (also on audiobook at Audible)

Quantum God Fractal Jesus by Andrew Fletcher

There's Nothing There, but Nothing is Really Something by Andy Fletcher

Answers to Questions People Are Actually Asking Book 1 by Andy Fletcher

All four found at www.lulu.com/spotlight/andyfletch.

God-ish books:

One Nation Over God: The Americanization of Christianity by DA Fletcher

Love Story by DA Fletcher

Both found at www.lulu.com/spotlight/andyfletch42.

The Authorized Autobiographical Writings:

Zombie Tacos from Hell: The Wit and Wisdom of Fletch

Found at www.lulu.com/spotlight/andyfletch42.

All the books are also found on Amazon and other fine, on-line book sites. Finding them in a bookstore (do they still have those?) is unlikely.

For film series or speaking engagements, contact me via my website—www.lifeuniverseverything.org and/or https://andyfletch3.wixsite.com/fletchblog.

Blogs: https://fletchblog42.blogspot.com/
https://onenationovergod647074800.wordpress.com/
https://fletchblog137.blogspot.com/And for all the science that seemed a bit vague to you, here're definitions and stuff:

GLOSSARY OF TERMS

Evolution: Darwinian evolution proposes that complex life forms arrived from simpler life forms via natural selection, that eventually species separated to become new species, that this all happened gradually over immense periods of time, that there is no direction to evolution other than survival, that the fittest to survive will survive and those which are not fit will be marginalized or cease to exist. Darwin did not know how creatures adapted to changing environments so as to be able to survive better.

Neo-Darwinian evolution arrived with the discovery of the gene and of genetic mutations. Random mutations in the genome thus became the single operative engine that drove and continues to drive evolutionary change. Random mutations on one gene in one organism provide for a change in behavior or morphology (the physical body) that coincidentally happens as it is needed to provide for better survivability in the face of this environmental challenge, and also happens to be able to be reproduced in subsequent generations. Thus a community of new and better creatures gradually arises, replacing the old, less adapted community.

There is no other engine to drive evolutionary change in traditional neo-Darwinian evolution, and there is no explanation for the arrival of life from within evolutionary theory.

There are other proposals in modern evolutionary theory that suggest that random mutation and natural selection play a less significant or even a minor role in providing for evolutionary changes. See the section entitle "Extra Bits" on page 183 for several possibilities in this regard.

Creationism: Comes in several flavors—young earth, old earth, Intelligent Design, and Fine-tuning. There is some overlap.

Young earth creationism: Everything was created in six days by God about 6000 years ago. The six-day creation cosmology comes from the Bible. The proposal that it happened 6000 years ago comes from someone called Bishop Ussher in the 1600s, who calculated the dates by adding up all the ages of the people in the Old Testament chronologies and subtracting backwards. Martin Luther and Johannes Kepler helped him with his calculations. More info on Ussher at www.law.umkc.edu/faculty/projects/ftrials/scopes/ussher.html .

Old earth creationism: Generally accepts an old earth and universe (4+ billion years for earth, 14.7 billion years for the universe) with a six-day creation cosmology. More info can be found at www.nwcreation.net/ageold.html and www.answersincreation.org.

Intelligent Design: Generally old-earth oriented; the major thesis is that biological organisms and systems appear too complex to have appeared without an outside design. Too many interrelated parts and pieces would have had to have arrived simultaneously in order to have provided for better survivability, and it doesn't seem possible for random mutation to provide for multiple mutations simultaneously that work so neatly together. The mousetrap from Lehigh's Michael Behe is the standard example. The best source of information on ID is found at the Discovery Institute's website, www.discovery.org .

Fine-tuning: First proposed by chemist Lawrence Henderson in 1913, extended by physicist Robert Dicke in 1961 and Fred Hoyle in 1984, modernized by John Gribbin and Martin Rees in 1989's Cosmic Coincidences, and supported in Christendom by Hugh Ross. Fine-tuning looks at the parameters of the existence of order, structure, and life itself to propose that it is physically and mathematically impossible for such an earth and universe to have arrived by accident. Within the secular physics community, fine-tuning is called the Anthropic Principle, proposed by physicist Brandon Carter in the '70s. It is controversial, not widely accepted, not universally rejected, and has been used to make critically important predictions in physics that have turned out to be accurate and true, specifically by Sir Fred Hoyle (carbon resonance) and Nobel Laureate Steven Weinberg (the cosmological constant). Both Hoyle and Weinberg were/are atheists.

Parameters of Fine-Tuning: There are dozens; you can find them on Hugh Ross' website, www.reasons.org.

Theory of Everything: The Holy Grail of Science. Physicists, cosmologists, and everyone concerned with the origin of the universe would love to find a theory that unites the existence of everything in one neat theory, preferably in one simple, beautiful elegant equation. There is precedent for it: Einstein's $e=mc2$ and $G_{\mu\nu} = 8\pi\, T_{\mu\nu}$ express with amazing elegance the most complex properties of nature in simple, beautiful forms, and it is the latter equation above that probably started the impetus in physics to look for the simple, elegant answer as the right one, and that when we finally are able to express the universe in a theory, it will be in a simple, elegant equation.

Big Bang: Bluntly, the only theory of the origin of the universe for which there is any evidence to be found in science and nature. Though Einstein originally rejected the idea that the universe had a starting point, preferring with the rest of the science community to believe that the universe was eternal and immutable, it was his General Theory (G$\mu\nu$ = 8π T$\mu\nu$) that first suggested that the universe may have had a beginning point. Eventually he and nearly everyone else was forced to concede that it was true, and finally to become, if not comfortable with the idea, at least resigned to it. Briefly, the universe itself, space and time, arrived in a tiny tiny fraction of a second about 13.7 billion years ago, everything and the potential for everything arriving in a blinding flash of pure physics. It was rejected by science because it sounded too much like the first chapter of Genesis, and eventually rejected by many Protestants because it didn't sound enough like the first chapter of Genesis. Both science and Protestant Christianity continue to hope that evidence will be found to disprove Big Bang. No luck so far.

The Singularity: the General Theory of Relativity seems to say that the universe expanded super-rapidly out of what turns out to be an infinitely compressed point of pure energy potential called the Singularity. Every Black Hole is thought to contain a singularity, a place where the laws of physics not only break down, but in the case of Big Bang, the laws of physics actually originated. Space and time cease to exist in Black Holes, and Space and Time had their origin in Big Bang, so at the Big Bang Singularity, Space, Time, and the laws of physics, the fundamental forces of nature did not yet exist. The universe itself started from a dimensionless point that did not exist in space or time. Stephen Hawking and Roger Penrose are credited with much of the theoretical information about singularities. There are now several competing theories, one of them Hawkings' himself, that indicate an on-going struggle with concept of singularities.

Quantum Mechanics: Also known as Quantum Theory and Quantum Physics, abbreviated by this book to QM. QM is the theory of the structure and behavior of atoms and subatomic particles. It arrived mostly in the first part of the 20th Century, and the more we discovered about the world of the very small, the more disturbing it became. The Quantum, micro-world is nothing like our large, macro-world, but everything is made of quantum particles.

Quantum particles can be in two (or many, or all) places at the same time, can travel immense distances instantaneously without having to cross any of the territory in between the arrival and departure points, don't have a real location in space or time, exist in a sort of quantum uncertainty where their realities have not yet been determined, and have realities that are only determined when a human attempts to observe them.

There are many upsetting elements to Quantum Theory, the most disturbing of which is the role of the outside observer. In a universe where to science humans seem to have no unique role at all to play, where we are no different than animals or anything else, for that matter, it seems to take a human observation of the quantum space in order for reality to exist. As much as nobody within quantum physics itself or the rest of humanity likes this, no one has ever been able to get rid of the need of the observer in creating reality at the quantum level, and since everything is made of quantum things, well, there you go.

If you'd like to do additional reading on QM (as though that would help), be sure to read about: The Two-Slit experiment (where one particle goes through two slits at the same time, and changes what it does depending upon whether or not you are looking at it); the Heisenberg Uncertainty Principle (which implies that a particle has neither a position nor a velocity until you look at it); Entangled Particles (which communicate instantly over vast distances), Quantum Tunneling (where particles skip from here to there without going anywhere in between), Schroedinger's Cat (where a cat is rendered neither alive nor dead by a particle that is not being observed), Quantum Computers (which are both on and off at the same time), and of course, the Copenhagen Effect (nothing ever happens without an observation). There is also Quantum Teleportation, Quantum Biology, Quantum Water, Zero-point energy, and the list goes on.

Here's what physicist Roger Penrose from Oxford says about Quantum Mechanics and, in passing, Relativity: "Quantum reality is strange in many ways. Individual quantum particles can, at one time, be in two different places —or three, or four, or spread out throughout some region, perhaps wiggling around like a wave. Indeed, the "reality" that quantum theory seems to be telling us to believe in is so far removed from what we are used to that many quantum theorists would tell us to abandon the very notion of reality when considering phenomena at the scale of particles, atoms or even molecules.

"This seems rather hard to take, especially when we are also told that quantum behaviour rules all phenomena, and that even large-scale objects, being built from quantum ingredients, are themselves subject to the same quantum rules. Where does quantum non-reality leave off and the physical reality that we actually seem to experience begin to take over? Present-day quantum theory has no satisfactory answer to this question. My own viewpoint concerning this—and there are many other viewpoints—is that present-day quantum theory is not quite right, and that as the objects under consideration get more massive then the principles of Einstein's general relativity begin to clash with those of quantum mechanics, and a notion of reality that is more in accordance with our experiences will begin to emerge. The reader should be warned, however: quantum mechanics as it stands has no accepted observational evidence against it, and all such modifications remain speculative. Moreover, even general relativity, involving as it does the idea of a curved space-time, itself diverges from the notions of reality we are used to.

"Whether we look at the universe at the quantum scale or across the vast distances over which the effects of general relativity become clear, then, the common-sense reality of chairs, tables and other material things would seem to dissolve away, to be replaced by a deeper reality inhabiting the world of mathematics"

Multiverse: An attempt by science to get around both Fine-Tuning and the need for an observer in QM, the Multiverse is a very large, perhaps infinitely large collection of other universes, within which our universe lies. There is no physical evidence for even one other universe, much less an infinitely large number of universes. But it remains an attractive option to many, as you will read. It is primarily derived as a concept from String Theory, for which there is also no evidence yet, Inflation, which has come under strong critical scrutiny of late, and Hugh Everett's Many Worlds interpretation of Quantum Mechanics, for which there is, again, no evidence.

Inflation: An unimaginably small bit of time (1035 of a second) in the original second of Big Bang when the universe expanded unimaginably rapidly, and then slowed down again. Inflation is necessary to explain critically important aspects of our universe, but there are new and significant arguments against it. From Alan Guth, Andrei Linde, and Paul Steinhardt.

Special and General Theories of Relativity: From Albert Einstein, the Special Theory in 1905 (along with four other earth-shattering papers), the General Theory in 1915.

The Special Theory redefined our Newtonian understanding of Space and Time, which change at high speeds (close to the speed of light), which are actually what the universe is made of (Space-Time), and which at the speed of light ceases to exist. That is, at the speed of light, all of time is experienced in a single instant, and all of space is compressed to two dimensions. "Space-time dilation" is the official name of what space-time does at high speeds.

The General Theory redefined our Newtonian understanding of Gravity with respect to Space-Time; that is, gravitational objects (galaxies, stars, planets, you and me and anything made of matter) cause gravity to exist and be experienced by warping and bending the fabric of space-time itself. Though we speak of gravity as a force, it's really geometry—space and time are bent like the surface of a trampoline when you stand on it.

Strong Force, Weak Force, Electromagnetic Force, Gravity: The theory of hot Big Bang Cosmology proposes that all four of the fundamental forces of nature (strong, weak, electromagnetic, and gravity) were bound together in the Theory of Everything (TOE), one force, and that as Big Bang progressed, each force crystallized out of the process. First, gravity separated itself from the TOE, leaving the Grand Unified Theory (GUT), which is the other three forces still combined. Second, the Strong Force popped out of GUT, leaving behind the Electroweak. Finally, electromagnetism and the weak force resolved into their separate selves. All of this happened in the first tiny parts of the original second. We have evidence that this is true for the Strong, Weak, and Electromagnetic Forces. We do not yet have evidence that Gravity was bound together with the GUT, the other three, because we don't yet have the technology to generate the energy levels needed to provide the evidence.

Each force has its own particles that "mediate" each, or cause each force to operate. The Strong Force is mediated by gluons. The Weak Force is mediated by W and Z bosons, whatever those are. Electromagnetism is mediated by photons and virtual photons (virtual particles are particles that are almost there, but not quite). Gravity is thought to be mediated by "gravitons," but little evidence has been found to document their existence yet. If gravitons don't exist, then physics is in a world of trouble.

The Strong Force is responsible for holding atomic nuclei together. Although it is very strong, it works over just the region that includes the nucleus in an atom. It provides small-scale structure that allows for matter to exist.

The Weak Force, or Weak Nuclear Interaction, is responsible for some forms of radioactive decay. The Weak Force has a major role to play in thermonuclear reactions, that is, the burning of stars, and thus a role to play in the arrival of the elements in the burning and death of stars.

The Electromagnetic Force holds protons and neutrons together with the nucleus to form atoms, and holds atoms together to form molecules. It also keeps solids from passing through each other via repulsion. Thus, it provides the everyday structure that we see around us and are a part of.

Gravity is the weakest of all the forces, 1033 times weaker than the next strongest force, but its reach is effectively infinite. Gravity is an interaction between things made of matter, which have mass, and space-time, and provides all of the large-scale structure in the universe. Everything made of matter bends space-time and therefore has gravity of its own, down to the smallest of atoms. We think of gravity as an attractive force field, in that things are attracted to one another via gravity. But gravity is in fact a geometrical warping of space-time by objects with mass.

Light year: the distance that light travels in space in one earth year. Light travels at just under 300,000 km/sec (just over 186,000 miles/.sec), which is more than 7 times around the earth at the equator in one second, and 8.5 minutes to the sun. So that's:

1,080,000,000 km/hour (1 billion km/hr)

25,920,000,000 km/day (almost 26 billion km/day) and

9,460,800,000,000 km/year (almost 9.5 trillion km/year, or almost 5.9 trillion miles per year)

A light year is not a unit of time—it's a unit of distance.

Entropy: the amount of disorder in a system. Entropy always increases in the universe, which means that the universe is always going from more ordered to less ordered. It's kinda like a house with teenagers. If you start with a pristine, brilliantly clean and ordered house, and insert some teenagers, well, you know what tends to happen pretty quickly. That's a metaphor for entropy—the house goes from ordered, neat and clean, to disordered, messy and dirty.

In a brand new house, just built, never lived in, you can tell by looking at it that it's new. No marks on the wall, no stains on the carpets, no dings on the door. If it's a used house that you are moving into after the previous owners moved out, they may have cleaned it really well, but there will still be little bits of evidence that it's not new. After you move in and live awhile, still more evidence that it's lived in. And if you have small children and pets, then the evidence begins to mount.

So it is with the universe. Our new universe was almost perfect (though it clearly didn't have to be, and we don't know why it was that way), and the older it got, the less perfect it became. That is, the new universe was almost perfectly the same temperature and density everywhere, but our present universe has really hot places and really cold places, really dense places and really empty places. It's older and has a lot more entropy.

The question is, with everything getting more disordered all the time, how do we end up with ordered things like galaxies, planetary systems, and you and me? The answer is, the laws of physics take the universe into more entropy, but along the way, those same laws cause things to clump together in bunches of disorder. You represent more disorder in the universe, but in a wonderful way, the laws of nature have produced the most ordered, complicated thing in the universe in that process—the human brain.

The Anthropic Principle: Physicist Brandon Carter in the '70s, preceded by evolutionary biologist Alfred Russel Wallace in 1904 and Robert Dicke in 1961, looked around at the parameters in the universe necessary for life to exist and decided that they were improbable enough that it bore investigating. His investigations caused him to announce in a symposium in 1973 that humans, or more specifically life in general, occupied a special place in space and time. From Wikipedia: "The phrase 'anthropic principle'" first appeared in Brandon Carter's contribution to a 1973 Kraków symposium honouring Copernicus's 500th birthday. Carter, a theoretical astrophysicist, articulated the Anthropic Principle in reaction to the Copernican Principle, which states that humans do not occupy a privileged position (in time as well as space) in the Universe. As Carter said: 'Although our situation is not necessarily central, it is inevitably privileged to some extent.' "

Wallace wrote in 1904, "Such a vast and complex universe as that which we know exists around us, may have been absolutely required ... in order to produce a world that should be precisely adapted in every detail for the orderly development of life culminating in man." Conditioned by biological factors ... [changes in the values of the fundamental constants of physics] would preclude the existence of man to consider the problem."

Since that time, it has become a controversial and yet hard-to-debunk principle in physics—the universe seems to have exactly the right set of scientific laws and realities to produce and support life.

One part of the AP that is not as often discussed is the need for an intelligent observer to cause reality to come into being—the Copenhagen, or standard Interpretation of Quantum Physics. Princeton Physicist John Wheeler was the leading proponent of what he called the Participatory AP.

Other parts include many constants and values found in nature that cause it to appear "fine-tuned" for life, and perhaps even for human life. For information on these constants and values, visit Hugh Ross' website, www.reasons.org/fine-tuning-life-universe.

Wikipedia gives a well-balanced exploration of the substance and controversial nature of the AP.

Dark Energy: We're not going to beat this to death with details. In 1998, it was discovered that the universe was expanding more rapidly than it should have been. It should have been slowing down, but it was speeding up. It's like this: you ride your bike or skateboard down a hill and then head up a hill on the other side. What should happen is that you will gradually slow down—maybe you reach the top, maybe you stop in the middle of the hill. Likewise, our universe, after the big push of Big Bang, should have been slowing down, but somehow, it wasn't. This is like you on your bike or skateboard in the middle of the hill you are going up, gradually slowing, until suddenly you speed up, accelerate. What does it take for that to happen? It takes energy. So the universe needs some extra energy boost to begin accelerating its expansion. So what do you call energy that you can't see? Well, it's dark, since we can't see it, and it's energy, so we'll call it Dark Energy. It has to be there, it has to be just over 74% of everything in the universe, but we can't find it and don't know what it is.

Dark Matter: We're not going to beat this to death with details. In 1933, a Swiss astronomer named Fritz Zwicky at Cal Tech discovered that galaxies were spinning around too fast to be able to be held together by the matter that was in them. In fact, you needed about five times as much matter as was visible in a galaxy to keep it from spinning itself apart. So the matter had to be there, but we couldn't see it. What do you call matter you can't see? Well, it's dark, and it's matter, so we'll call it Dark Matter. Nobody paid any attention to Fritz because he was really irritating and hard to get along with. In 1975, astronomer Vera Rubin announced similar observations, and after a lot of controversy, she was believed. Dark matter is just under 22% of everything in the universe, but we can't find it and don't know what it is.

What's left—ordinary matter and stellar gas. Stellar gas is about 3.6% of what's in the universe, so ordinary ("baryonic") matter, all the stuff that you and I are made of and that we can see around us on earth and in the night sky, is about .4% of everything that exists in the universe. Now that's depressing. And, in the latest news, about half of that is now missing. Swell.

www.ingramcontent.com/pod-product-compliance
Lightning Source LLC
Chambersburg PA
CBHW060851170526
45158CB00001B/313